# I AM THESE TRUTHS

# I AM THESE TRUTHS

### A MEMOIR OF IDENTITY, JUSTICE, AND LIVING BETWEEN WORLDS

## SUNNY HOSTIN

#### WITH CHARISSE JONES

HarperOne
*An Imprint of* HarperCollins*Publishers*

HarperOne

FIRST EDITION

Designed by Terry McGrath

Library of Congress Cataloging-in-Publication Data has been applied for.

ISBN 978-0-06-295082-6

20 21 22 23 24   LSC   10 9 8 7 6 5 4 3 2

*Mom and Dad—I am because you sacrificed everything. Thank you.*
*Manny—I am because you have always supported my dreams. Thank you.*
*Gabriel and Paloma—You are my greatest blessings. I am because you are.*
*I love and adore you. Always.*

# CONTENTS

# FOREWORD

Just as my final edits on this book were due to my publisher, the standards and practices group at ABC News sent me pages of changes they wanted me to make to my story. It's not uncommon for an employer, especially of a journalist, to vet a manuscript for accuracy or even tone. The process had been taking longer than I expected; ABC explained that breaking news required their immediate attention. Of course it did: after George Floyd was killed by Minneapolis Police, hundreds of thousands of people around the world marched and demonstrated to end police violence against Black people. As a mother of a Black teenaged son about to go to college, I could wait. Covering these demonstrations required all hands on deck and of course my publisher allowed me a few extra days to resolve my employer's requests. While I was grateful that ABC News caught a few factual errors that would have embarrassed me, they were also asking me to delete parts of my story that might cast ABC in an unfavorable light. Deleting those passages didn't feel right to me—they were all true, and they were some of the battle scars of my experience.

My television agent and my book agent emailed me to express confusion that a news organization would try to censor a Puerto Rican, African American woman's story while they were covering global demonstrations demanding racial equity. One of them even calculated the percentages of people of color on the executive boards at Disney, ABC Entertainment, and ABC News—according to him those figures ranged from 7 to 12 percent. I asked my attorneys to intervene and thankfully ABC relented. I didn't want to believe that racism played a part in their revision requests—we were just dotting some i's and crossing some t's, right? Then, on Friday, June 12th, I got a text from a reporter.

Any journalist will tell you that they never want to become part of the story. But suddenly I found myself in that odd and unenviable position. The reporter explained that he was writing a story that would be published shortly, and that as a courtesy he wanted to give me a heads-up that I would be mentioned. I immediately thought this was yet another article about *The View*. The show was always the subject of media fodder. The summer before I was accused of being a leak of negative stories about the show. The accusation was hurtful and wholly untrue and I was still angry about it. Then the reporter explained that his story was not about *The View*, but rather about a series of racist comments made by a senior talent executive at ABC News. It was alleged that the executive had called me "low rent"; had exclaimed during contract negotiations for Robin Roberts, the anchor of the network's flagship morning show, *Good Morning America*, that she was not being asked "to pick cotton"; and that my former ABC News office mate, Kendis Gibson, currently an MSNBC anchor, one of the nicest people I've met, would not be paid as much as ABC paid for toilet paper; and finally that Mara Schiavocampo, a former ABC colleague, had entered into a non-disclosure agreement stemming told my children, my son was incredulous: "That is the farthest thing from the truth, Mom. Who would say that about you of all people?" My daughter was pensive. Then, she just said, "Why?"

I really did not have an answer to the "Why?" I was assured that an independent investigation would be conducted and if the allegations were true, proper action would be taken and that there would be a "recalibration" for me. I'm not really sure what that means. *The View* gave me the space I needed the Monday after the story broke to speak my truth. While it was painful for this tough girl from the Bronx to show any vulnerability, I did. I explained what had been said about Robin Roberts, Kendis Gibson, Mara Schiavocampo, and me. And I said, "If true, this tells me that systemic racism touches everything and everyone in our society. Regardless of social stature, no one is immune. It is the type of racism that Black people deal with every single day, and it has to stop."

I fear for the lives of my teenaged children each and every day. Their lives matter. Black. Lives. Matter. But we can't just say it—our country needs to show it, to believe it, to mean it. Black people are arrested and brutalized by police at a stunningly disproportionate rate around the country. Black people are incarcerated at more than five times the rate of white people. The deaths from this devastating global pandemic are disproportionately high for Black communities.

As I send these words to press, we are experiencing an important time in our country's history. A paradigm shift. A movement, not just a moment. Our world is on fire. It is demanding that voices be heard. I envision a world where our hearts and minds cling to a promise of love, humanity, compassion, and most importantly, equality. I envision a world that seeks to reign above pettiness, divisiveness, and hatred. I envision a world not of fences or barriers, but a world where these from racial discrimination and could not comment on her alleged treatment at ABC. This executive not only managed ABC's diversity and inclusion efforts, but she was directly responsible for cultivating my career, negotiating my contracts and providing network opportunities. I was floored. I felt incredibly sad, but I also felt relief. Many of the experiences I've had at ABC, including several described in these pages that standards and practices at first asked me to delete—well, if the allegations were true, all of the dots were connected. My suspicions that I was treated worse than my white colleagues—the fears that I tried to talk myself out of many times—maybe they were true. Had my employer, my home away from home, devalued, dismissed, and underpaid me because of my race? I had just read emails from them directing me to erase evidence of such treatment from my story.

And if I'm being honest, I wasn't even angry. I was deeply, profoundly shaken and saddened.

I had been honest on air about my humble beginnings, and about growing up in public housing. I thought it was important for anyone watching in similar positions to know that being a national talk show host was within their reach. And now, my roots were being disparaged. I was seen as not worthy, classless, cheap, trashy, low rent. I cried. Silently alone. The next day, the article was published, and went viral. The executive was placed on administrative leave. I received hundreds of calls and emails. I received calls from my cohosts, colleagues, former colleagues, the top brass at ABC and Disney. Thousands of social media comments. Most of support, but also many calling me a "low rent bitch." I hoped that my children wouldn't find out, but I knew that this had gotten too big. I finally told my parents and my husband. My parents were angrier than I had seen them in a very long time. My husband held my hand. When I finally protests lead to equality and humanity for each person regardless of skin color.

Over and over again, we have seen the weaponization of race against Black people in our country. Driving while Black, Jogging while Black, Barbecuing while Black, Shopping while Black, Bird-watching while Black: Living while Black may lead to confrontation, police intervention, death. Where the innocent are assumed guilty, because of the color of their skin. I hope to see a reckoning. Real introspection combined with real action. Because what is going on from the street, to the classroom, to housing, to financial credit, to hospitals, to the newsroom—it has to stop. Now. We need a recalibration.

I do not believe that I made it from public housing in the Bronx to these privileged television studios because I am exceptional. I believe that anyone who is treated equitably can attain a future they want. I do not believe that my success shields me from racism and discrimination. I do believe that we can, right now, make changes to create a level playing field for all. We all have to work for it together. I Am These Truths.

Sunny Hostin
June 15, 2020

# I AM THESE TRUTHS

# CHAPTER ONE

# THE BOOGIE DOWN

In a decade of change, 1968 marked a turning point. Americans were reeling in the wake of the assassinations of the Reverend Martin Luther King Jr., murdered on the balcony of a Memphis motel, and Senator Robert F. Kennedy, killed moments after winning the California primary. Police savagely beat demonstrators protesting the Vietnam War during the Democratic National Convention in Chicago, and the nonviolent underpinnings of the civil rights movement were giving way to the fist-in-the-air, by-any-means-necessary defiance of black power. Some wondered if our broken nation could ever be put back together, a question that many Americans are asking themselves today.

It was in that year that I was born to teenage parents, both of whom were activists in their own right, protesting segregated housing and schools that existed not just down south but right in their own backyard in New York. It was as if the need to fight for social justice was in my parents' blood and embedded in my DNA.

While my parents' quest for equity definitively helped to shape the woman that I am, my beginnings were less than auspicious. My parents were high school sweethearts with towering dreams. They were plotting their own course to help save the world, or at the very least etching a plan to do their small part to change it. But while they would eventually make their mark, an unexpected pregnancy forced them to travel down a far more jagged path.

My Puerto Rican mother, Rosa Adelaida Beza, grew up on the Lower East Side in a tenement apartment with a bathtub in the kitchen where you would wash both your dishes and your hair and body. My mom lived with her mother, Virginia; her older sister, Carmen; and her baby sister, Inez. Always enamored of black culture, she felt close to it, of it, and wanted to do her part in combating what she felt was senseless racial injustice she witnessed in the news and in her own neighborhood. She read all about Malcolm X, the Black Panthers, and the FALN, and she secretly dreamed of one day joining or even starting a social justice organization.

By the time Mom turned seventeen, she longed to be the first in her family to go to college. Perhaps she'd participate in sit-ins at a college out west or witness voting rights reforms down south. She wasn't sure where she would study, or where her next picket line might be, but she was eager to find out.

Meanwhile, my eighteen-year-old African American father, Willie Moses Cummings, had lofty goals of his own. Tall and lean with a picked-out Afro, Dad was a math whiz who had been a track star in junior high school but ultimately had to give up running so he could work after school to support himself and help pay the house bills. When he wasn't working or in class, he could be found curled up in a corner of his family's Harlem apartment reading Martin Luther King Jr.'s "Letter from a Birmingham Jail," or *The Prophet*, by the

Lebanese American writer Kahlil Gibran. Introspective and studious, Dad recognized the merits of boycotts and demonstrations and also wanted to be a part of the movement. He had dreams of running track in college and yearned to use his intelligence and compassion to help others by becoming a physician.

My father was a high school junior when he first caught a glimpse of my mother. She hung out in the same group as his rambunctious younger brother, Eddie James. He says that when he laid eyes on the pretty Puerto Rican girl with the activist spirit, it was love at first sight. Mom was not so instantly smitten—she thought he was a little "corny" and played it too safe. She remembers that her affection grew over time, budding in rushed conversations by her locker and blossoming after sneaking out to a few Harlem basement parties. Dad kept chasing, and finally Mom let him catch up.

But there would be no prom for them, no corsages or goofy pictures snapped outside a limousine. My mom wouldn't walk across Seward Park High School's scuffed stage to receive the diploma that would be her ticket to take flight. The first time my parents were intimate, Mom got pregnant, and with that, her and my father's childhood came to an abrupt end.

Mom was able to hide it for seven months, swaddling her growing belly under baggy sweaters and coats and forgoing prenatal care. But while that camouflage worked in winter, it wasn't possible or practical in the broiling heat of a New York summer. Finally, my grandmother Nannie Virginia discovered that she had a grandchild on the way.

"Well," Nannie said, probably after making the sign of the cross and picking her jaw up off the floor, "I guess we're going to have a wedding."

My teenage parents tied the knot on October 5, 1968. My mother wore an incredibly fashionable mini wedding dress, with her red blond hair styled in a Mia Farrow/Twiggy pixie cut. I always thought her wedding bouquet was way too big—she explained later it was to cover her tummy, lest people realized she was pregnant. I came into the world fifteen days later, in Manhattan's Beth Israel Hospital, on October 20, 1968. Dad put away his medical school dreams for good, enrolling in a technical school instead. And Mom did leave the Lower East Side, but not for college. She dropped out of eleventh grade, got her general education diploma, and headed out to live at my paternal grandmother's apartment in the South Bronx.

My journey began way uptown. I took my first steps in the hallways of Nannie Mary's home, a high-rise at 1889 Sedgwick Avenue, smack-dab in the mythical heart of the Boogie Down Bronx.

* * *

In her own way, my father's mother was an immigrant, even a refugee, though she was fleeing terror in one region of the country of her birth to seek safety in another. She and her family were among the flood of African Americans who escaped the bigotry and cruelty of the Jim Crow South to unearth opportunities elsewhere. The push, to Chicago, to New York—to virtually anywhere so long as it was north of the Mason-Dixon Line—began in the twenties, near the dawn of the resurgence of the Ku Klux Klan, and continued until the 1960s, when the movement to gain black people the rights they should already have had was in full throttle. When it was over, roughly five million men, women, and children had uprooted their lives to carve out new homes and destinies in urban hubs all over the United States.

Nannie Mary's father had been a sharecropper with his own small piece of land in Georgia, but despite laboring long hours, it was a constant struggle to provide for his ten children and his wife, Lilly Mae, who also helped tend the land. Life in Georgia wasn't easy, and eventually, after giving birth to two sons, and watching relative after relative move North, Nannie Mary finally decided to take a chance too, making the journey with my dad and his brother in tow. They moved around a lot, never quite being able to make ends meet, and finally settled in the heart of Harlem. Harlem shaped my dad, washing off a lot of his Southern ways, although a few remained. I don't ever remember my father having a thick Southern accent like my uncles and cousins, but he made the best grits for Sunday breakfast and had a penchant for cowboy boots.

As Harlem became more and more expensive, Nannie Mary finally settled in the Bronx, her home becoming the family hearth, where everyone would congregate and celebrate just being together. Nannie Mary was small in stature (barely five foot) but tall in personality. Some thought she was too abrupt, but that's what I loved about her. She was a truth teller. A sharp dresser and a great cook.

There was always a pot bubbling on the stove, the smell of oxtails or collard greens wafting into the hallways and beckoning you inside before you even opened the front door. And there was nothing like Thanksgiving, with the turkey, corn bread, black-eyed peas, rice, sweet potato pie, and banana pudding (with the Nilla Wafers, of course). The young ones would pretend to know how to play cards, amid all the yelling as the grown folks listened to music, drank, danced, and played bid whist.

My cousins and I would play hide-and-go-seek, infuriating my grandmother when we burrowed inside her closet or the bathtub, and we'd stay up long past the time we were supposed to be asleep, spinning stories and cracking jokes. The three of us, Sean, Tyvee, and I, remain incredibly close to this day.

Even when our extended family wasn't visiting, which was most of the time, Nannie Mary's apartment was a crowded space. I slept in one of the two bedrooms with my mother and father, while Nannie Mary shared the other bedroom with her husband, Doctor Dash (yes, Doctor was his real name), whom we all called Doc.

I adored him. He had a solid job working for the New York Department of Sanitation. He would lift and tote overflowing garbage cans all day, then dutifully bring his paycheck home to my grandmother every two weeks, setting aside a couple of dollars to slip to my cousins and me on the side. Doc was as sweet as the Now and Laters that he always carried in his pocket as a treat for me.

I can't remember him ever raising his voice at anybody, not when he tripped over one of my dolls forgotten and left in the middle of the kitchen floor; not at my uncle, who would sometimes come over in the middle of the night, alcohol on his breath, disrupting Doc's much-needed sleep; not even at the neighbors down the hall, whose trash somehow never made it into the incinerator chute, cluttering the hallway all the neighbors had to share.

But Doc was a bit of a character, more like a big kid than an elder. I remember one afternoon, it was just him and me. He would do basically anything to make me laugh, and that day he linked his broad hands with mine and began to swing me around. I could barely catch my breath, I was giggling so hard, the plastic-covered burgundy couch, the collection of family photographs, and the painting of a golden-haired, blue-eyed Jesus all whizzing by in a blur.

I'd wanted him to swing me. But of course, he shouldn't have.

First I heard a pop. And then I felt a searing pain. I stopped giggling and started crying. Doc knew he was in trouble.

It turned out that he had dislocated my shoulder. That, however, wouldn't be diagnosed for hours. Despite his name, Doc didn't have a clue what to do, and he was panicking because he knew that he was going to get an earful from Nannie Mary, Daddy, and, probably most of all, my mother.

He propped me on the couch like a bent rag doll. "Don't tell anybody that we were playing," Doc said, glancing furtively over his shoulder, awaiting the moment of doom when one of the other grown-ups in our family got home. "If you sit still, after a while, you'll feel better. Okay?"

"Okay," I said softly. I was normally a ball of motion, running the short length of those four rooms like it was the rolling lawn at Van Cortlandt Park. But I was going to do my best not to move too much because every time I did, I felt a shard of pain ripple down my arm.

I sat there for what seemed like hours, the sunlight that trickled into the living room fading from bright yellow, to tangerine, and finally to a dull amber. Doc clicked on the TV to try to keep me entertained, cracked open a can of beer to try to keep himself distracted or at least somewhat calm, and nervously paced the floor.

Finally, Mom came home. I can't recall exactly what happened next, but I remember some yelling after Mom gave me a hug and I let out a piercing wail. I was whisked to the hospital and got my shoulder popped back in place. But it was a long time before I was allowed to hang out with Doc alone again. And I can tell you I missed him.

* * *

As wild as that moment was, it wasn't all that unusual. The truth was that there was a lot of madness in Nannie Mary's house. Doc was a drinker, who made his own moonshine on the balcony, and he wasn't the only one of my relatives who had a love affair with liquor. On my mom's side, Nannie Virginia's fourth husband, Antonio, whom we all called Tony, often smelled of cigarette smoke and beer. And while Nannie Mary eventually became a member of the Pentecostal religious tradition, swearing off alcohol forever, before she got saved, she could sit in her armchair and have a few.

The constant stream of chatter and liquor-laced laughter made life at Nannie Mary's feel like a never-ending, ever-ready-to-pop-off party. But it made life combustible too. Loud talk and playful boasts about something as innocuous as why the Yankees lost their last game, or a bet on how many days it would take the city to steer its snowplows to the South Bronx after a storm, would suddenly spill over into a heated argument about a long-ago slight that had never been forgiven. My great-uncles or older, distant cousins would bolt out of their chairs, fists clenched, faces bobbing close enough to kiss. They rarely came to blows, but many times the rest of the family thought that they might.

In Nannie Mary's apartment, there always seemed to be something ready to ignite, whether it was tensions that mounted from counting every penny only to still come up short, or the time Doc actually set the living room curtains on fire—probably because he was stumbling drunkenly with a smoldering cigarette.

Decades later, my husband would remark on how I was the kind of person you wanted around during an emergency because while others become frantic, fear overwhelming their thoughts and actions, I tend to slow down. I'm able to be still, to stand apart, to mentally remove myself from the situation so that a solution comes to me in sharp relief. It's a behavior that came in handy when I had to tend to my son after he pulled a hamstring and lay on the field writhing in pain while playing high school football, or when I needed to figure out what to do when a witness failed to show up to testify in a case I was arguing in court. I honed that steadiness in the Boogie Down Bronx.

I learned to be calm in the midst of the storm swirling around me, to not become part of the turmoil. I learned to not panic as orange flames devoured Nannie Mary's curtains; to focus on the words on the pages of *The Lion, the Witch and the Wardrobe* and not the curse words being flung back and forth outside and inside our home.

In the midst of all that confusion sat my mother like a fish out of water. She'd wanted to leave Nannie Mary's home from the time she arrived. Her own family, ensconced on the Lower East Side, felt a million miles away. And she never got along that well with my father's mother, who was set in her ways and never let Mom forget that this was *her* house. She acted like she was doing my mother a favor by letting her live there. Most of all I believe Mom worried about me, growing up in such a tumultuous home.

Even as a very young girl, I could see how unhappy Mom was. It was etched in her furrowed brow when Nannie Mary started complaining. I could see it when she would sit in a corner drinking her own stiff drink while watching the madness unfold. I could feel it when she retreated to her room when Doc and Uncle Ed got to drinking, and I could hear it when she spoke to my father about the need for our family to find our own place. She wanted more. She wanted better. She wanted out of there.

And so, when I was about three or four years old, my parents found a fifth-floor apartment on Morris Avenue, near the Grand Concourse, one of the major avenues slicing through the Bronx.

\* \* \*

It was the early 1970s, and the hippie ethos of peace and love permeated the fashions we wore, the music we danced to, and even the décor in our new home. Instead of opening and shutting doors, Mom hung long strands of beads that would click and shimmer as we walked by. And for the first time, I had my own room, and Mom and Dad had theirs.

Now that our family was living on its own, my parents had to stretch their dollars as far as they could go, boiling water to warm the air on the many days that the heat was cut off and creating veritable feasts out of chicken hearts, bad cuts of steak, and Spam. My mother made most of my clothes, buying bolts of colorful cloth and running the sewing machine in her bedroom at all hours of the night to turn them into rompers, dresses, and skirts. But I never felt deprived. We'd go on a picnic in the park, or a trip to the Cloisters, the cluster of art-filled medieval buildings imported from Europe and transplanted to Fort Tryon Park, and I'd pretend to be asleep in the back seat of my parents' gray Monte Carlo so Dad would tenderly lift me and carry me up the four flights of stairs to our top-floor apartment. We were happy.

But Mom and Dad wanted a home that didn't require jerry-rigging space heaters or boiling water on a stove to stay warm. Soon enough, we moved again, this time to the St. Mary's Park Houses on West-chester Avenue.

There had been a waiting list to get in. St. Mary's had an elevator and a playground. And it was near St. Anselm, which had a reputa-tion as one of the better Catholic schools in our corner of the Bronx. It was supposed to be middle-income housing.

I recently drove by there, and I have to admit it looked pretty shabby, a cluster of weathered buildings surrounding a bleak open space that was more parched and brown than bucolic and green. But at the time, St. Mary's seemed like a bit of luxury, like if we hadn't exactly arrived, at least we were on our way. Sure, the elevator smelled like pee. And when it wasn't working, which was more days than I could count, we had to walk up twelve flights instead of the four we'd climbed before. Now I had to endure every step because I was getting taller and heavier, and my possum act was no longer good enough to fool Dad into carrying me up all those stairs.

Sometimes the lights were out, and it was scary in the darkness. But once we walked into our five-room apartment, and I could see the treetops of St. Mary's Park from the living room window, all of those inconveniences felt small, inconsequential, and they were quickly forgotten.

My mother put her community-organizing skills to work, forming our building's first tenants' association, and our neighbors would take turns going on patrol. At Christmastime, everyone would decorate their doors with shiny wrapping paper and bows, and there were contests on each floor to judge who'd decorated the best. On Halloween, all of us kids would go trick-or-treating, ringing all the doorbells from the bottom of the high-rise all the way to the top, getting more candy than a pillowcase or jack-o'-lantern bucket could hold.

I once asked my father why we always lived at the tip-top of every building that we called home.

"Because that's the penthouse," he said with a sly smile. "Everyone wants to live there. That's the best spot to be."

Even as a little girl who hung on her daddy's every word, I'm not sure that I believed that story. But I always appreciated the loving spirit that made my father say it, his urge to always make me feel special, and the need to make the best of what we had, even as we strived for so much more. I knew he always wanted me to have not only better than he had. He wanted me to have the best.

We still lived near Nannie Mary, who I saw nearly every day. And there was a corner store I loved going to with Doc, Uncle Ed, my cousins, and all my friends.

The owner sat behind a shield of glass, and we'd slip change in a tray beneath it. He'd flip it around like a bulletproof lazy Susan, and Chick-O-Sticks and Mary Janes would magically appear. I'd tear off the wrapping, pop them in my mouth, and savor all that sweetness.

In the summertime, I'd cool off from the heat in geysers unleashed from illegally unplugged fire hydrants. And on weekends, Mom and I would shop on Fordham Road. The stuff being hawked under the rainbow-colored awnings wasn't the highest quality, but as far as I was concerned, we might as well have been strolling along Beverly Hills' Rodeo Drive. Vendors sold *piraguas* (I always got *tamarindo*) and *helado de coco*, while the sidewalks hummed from the crowds jostling for space and attention. It was like I got to live inside a constant carnival. How lucky was I?

Still, reality sometimes intruded on my rose-tinted gaze. The South Bronx was a web of some of the poorest neighborhoods in New York City, where one in four boys and girls struggled with asthma, and fights and shootings were as common as the blazes set by landlords trying to cash out of worthless buildings. People, turned off by the blight, depressed by the poverty, scared of the crime, left the Bronx in droves.

My path to and from school was strewn with broken glass. And though I never felt particularly poor, I knew that money was often tight. At Nannie Mary's, there seemed to always be murmured conversations about pooling money to pay the light bill, to buy a birthday gift, or to repair a broken-down car.

I also knew that no matter where we moved in the South Bronx, Mom felt that we could find better and do better back in her home borough of Manhattan. It became a drumbeat when we visited Nannie Virginia or Aunt Carmen.

"You've got to get out of the Bronx," Nannie Virginia would say.

"I know," Mom would say with a sigh, that pensive look I saw so often at Nannie Mary's returning to her face. "We've just got to get our money together and find a decent place."

Mom and Dad were on a mission, to get their degrees, to get a nicer apartment in a safer neighborhood, and to create a safe perch for the baby girl they hadn't expected but who, now that I was here, they would do anything for.

* * *

It had become clear that while the St. Mary's Houses were supposed to be a come up, being there actually just shuttled us steps closer to the violence and despair that had always lurked right outside our front door. I guess it was inevitable that one day the violence that regularly erupted beyond our walls would step up and strike me in the face.

One day, I was hanging out with my best friend, Angelique. We were playing hopscotch, calling each other out for accidentally slipping a sneaker across a chalk-drawn line, when a bunch of other kids in the neighborhood began running toward us, screaming something we couldn't make out. As they got closer, their words became clear, the battle cry of a nightmare.

"Angelique's father got shot!" they yelled. "Angelique's father got shot!"

We ran toward the danger as the other kids ran away. By the time we got to where Angelique's father had fallen, the police had arrived, along with an ambulance. Angelique was sobbing. I clutched her hand before one of her relatives, who suddenly appeared in the gathering crowd, grabbed us both and rushed us away.

If I ever knew what led to Mr. Moreno being shot, the reasons have long faded from my memory, and Angelique—who is still a friend—gets a cloud over her face if anyone tries to talk about it. I do know that it wasn't the first time either Angelique or I had heard our schoolmates scream that someone had been hurt. It wasn't unusual to see a crowd run away from gunshots or toward the spectacle of a fight. But it was the first time that the person they were screaming about was someone I knew well.

Angelique's father died, inexorably changing the emotional arc of her life. And as awful as that was, soon, for me, violence would hit even closer to home.

* * *

My uncle Ed was my father's baby brother, tall, brown, with a smile that made you feel like you were the most intriguing person in the world. Whenever I hung out with Uncle Ed, I felt like I was in the mix, like somehow whatever excitement was percolating, I was at the center of it with the man everybody wanted to be around.

When Uncle Ed wasn't drinking or high, there was no one kinder.

Unfortunately, I seldom saw him sober. All that charm and charisma provided cover to a troubled soul. Uncle Ed was smart, but he suffered from addictions and could never keep a job. He constantly needed to borrow money, from Nannie Mary, from Doc, from Dad, or from his girlfriend du jour.

Dad had always been his protector, getting him out of scrapes with boys whose sisters had been left crying after Uncle Ed cheated on them, or relatives who were tired of his promising to come by and help with a task, then failing to show up. But there came a day when Uncle Ed got caught up in something that Daddy couldn't save him from.

I think I was about seven years old, and some details of that day have gone fuzzy in my mind. What is clear is that I was hanging out with Uncle Ed. There was a woman there as well. And what came next seemed to unfurl in slow motion. We were in an apartment, though I don't remember whose. I know that there were red couches covered in plastic and beads like those that hung in my parents' first apartment.

Ever the womanizer, I guess Uncle Ed was dating someone else's wife or girlfriend. We were sitting around when suddenly a guy burst through the door. He and Uncle Ed began to argue, and then, so fast that you could barely see the blade, the man pulled out a knife and stabbed my uncle several times.

I didn't get swept up in the madness. I didn't lose my mind in the whirlwind. I slowed down and tried to figure out how I could help my uncle.

Uncle Ed ran into the bathroom, and I ran behind him and closed and locked the door. I grabbed the roll of toilet paper to try to stop the bleeding, and amid the stream of blood seeping onto the black-and-white tiled floor, I glimpsed splotches of yellow—probably my uncle's intestines.

As I tried my best to apply first aid, while Uncle Ed clutched his stomach and yelped in pain, my grandmother and mother appeared, wailing in desperation. An ambulance was called, and Uncle Ed was taken to the hospital. He nearly died, but despite the situation being touch and go for several days, he survived.

Years later, I would recall that day in a conversation with my father, who was amazed that I remembered it.

"You were so young," he said. "We never talked about it, so I thought—we hoped—you'd forgotten."

But how can you ever forget something like that? The trauma of it lurks in the core of your being, impacting your actions and your reactions in ways that you may not first realize and may never fully understand. It made me question how someone could be so foolish—"You date someone else's wife? What do you expect?"—but because I so loved my uncle, it also imbued me with empathy and the ability to see that there were multiple sides to every story.

I understood implicitly that anyone could make a foolish mistake, that you could feel the urge to make a quick buck, or fall in love with the wrong person, and yet that did not justify nearly losing your life. It was the beginning of my visceral understanding of the causes, and the costs, of violence.

And what I also know now is that what happened to Uncle Ed, and Angelique's father, was not something unique to the Bronx or to the circle of people I loved. The pressures of poverty, of police harassment, of low expectations, can certainly exacerbate tensions, can perhaps make a person on the edge snap a little more quickly. But addiction, feuding, and infidelity are as prevalent in Hollywood and Iowa as they are in the Bronx. It's just that for those who are wealthy or white, a whole group isn't defined by the unfortunate actions of a few. They are more often afforded the benefit of the doubt. Context is given to their actions, even if they don't offer it themselves, because they have the money or the privileged status to demand it.

The tragedy that engulfed Uncle Ed marked the beginning of the end of calling the South Bronx home. My parents were pillars of stability who fought to get me into the best schools and to engage my imagination with trips to museums and the planetarium. They worked in unison to protect and propel me. But beyond their united front were outside forces creating instances of danger and violence. I'd be at Nannie Mary's, and curtains would catch on fire. I'd spend the day happily hanging out with my uncle, only to watch helplessly as he got stabbed before my eyes.

So much chaos, so much turmoil, so much my mother and father could not control.

"I'm done," my mother said after my father picked me up and brought me home as Uncle Ed lay in the hospital. By summer, we were in Manhattan at last.

But long after we'd packed our boxes and moved away, the South Bronx remained as dear to me as the alphabet soup of streets on the Lower East Side where I would spend my middle and high school years. It was where Nannie Mary and Doc and Angelique continued to live. The love that I felt within my relatives' homes, and the desperation that I glimpsed inside and beyond them, forever fueled me, influencing the stories that I wanted to tell and dictating the people for whom I wanted to fight as a social justice advocate, as a lawyer, and as a journalist. But now, I was going to lay my head miles away.

The day we moved, my parents rented a U-Haul truck, but we hardly needed it. We were leaving much of our furniture behind. The worn brown couch where Dad and I would sit and watch *The Flintstones* and reruns of *I Love Lucy*, the chipped table where Mom would rest her cup of café con leche in the morning, the rug worn through from my dancing to Stevie Wonder and Rufus and Chaka Khan, all stayed there.

We were also hopefully leaving behind the memories of blood

spilling onto a black-and-white tiled bathroom floor, voices of strangers yelling that a friend's daddy might be dead, and all those sharp edges of shattered glass and broken lives that I passed on the way to and from home.

Mom, Dad, and I rode the elevator down in the St. Mary's Houses for the last time. My father revved the motor of our Monte Carlo, and we slowly headed down Westchester Avenue. We didn't look back.

CHAPTER TWO

# SCHOOL DAYS

What's up, traitor?"

That was how Freddy, a little boy I'd known for as long as I could remember, greeted me one Sunday afternoon. I was giddy to be back in the South Bronx for one of the first times since my parents and I had moved to Manhattan, and I was practically skipping down the sidewalk with my best friends, Tanya and Angelique, happy to see so many familiar faces, including Fred's. But his taunt stopped me in my tracks.

"Traitor?" I thought to myself. "That was kinda mean. Why'd he say that?"

I mumbled a shaky hi and then turned back to chatting with my girlfriends. But I was startled. Unfortunately, it wouldn't be the last time one of my old playmates hurled an insult at me when I returned home every other weekend so that I could spend time with Nannie Mary and the rest of my father's side of the family.

I had been used to trash talking on the playground. It was practically a sport in the South Bronx. Two kids would be dissing each other back and forth, but when I joined in, they'd drop whatever they were arguing about and tell me to mind my own business.

"Why are you getting in it?" one would yell. "You think you're so big now because you live in Manhattan!"

Or I'd mention a field trip I took with my new classmates to see Revolutionary War landmarks near Wall Street, only to catch Maria, an old double Dutch buddy, rolling her eyes.

"So what?" she'd say, her eyes flashing. "You're not better than us just because you go to school in Manhattan!"

At the time, the only thing clear to me was how much those words hurt. But I eventually understood that my move to Manhattan, and the shuttling to and from the Bronx that would consume my Saturdays and Sundays, was the start of something that would become a defining theme of my life.

It marked the beginning of having to straddle multiple worlds. A balancing act that would become as natural and necessary as juggling motherhood and a career, of loving new and old friends, but having to recognize that they wouldn't always fit or be welcome in the places where I would go—and vice versa.

Those playground jabs were the first hints that I would need to learn to submerge parts of myself depending on the rhythms of the room; that I would need to leave some feelings and experiences unspoken so as not to create even more distance between myself and the people in my circle. It was the beginning of feeling, despite the fierce embrace of my family, that sometimes I would just have to figure it all out on my own.

I knew in my heart that I hadn't abandoned my old friends, that I never would. After all, I was standing right there, playing the same old games on the same cracked sidewalks, within earshot of every joke and piece of gossip. I was there so often that it felt like I'd never really left.

But as much as I felt connected by the umbilical cord to the South Bronx, I also had no regrets about my new home. Our new apartment was so much more beautiful, my new school so much nicer, my new neighborhood so much cleaner and safer than what my parents and I had left behind. My parents had made the right choice. And from the time I was very young, they had instilled in me the belief that I should never be ashamed of wanting better and never pass up the chance to grab ahold of an opportunity.

* * *

To be honest, though, the gulf between myself and many of my old friends began to emerge long before I ever left the Bronx. In the apartments of many of my friends, the TV was usually on or the music blared. Education was left to the hours between eight and three, when kids were in school.

But in my home, education was at the forefront of the conversations we had and the activities that we did. I spent afternoons playing jacks and jump rope with Angelique, Lisette, Celeste, and Tanya, but when it was just me and my parents, books were the primary form of recreation.

I can't remember not knowing how to read, but I believe the instruction began in earnest when I was about four. My mother didn't want me in day care, so she put off getting a job so that she could stay home with me. In the evenings, she attended Bronx Community College, studying to be a teacher, and our days were filled with

primers and lessons as we sat side by side, Mom mastering the principles of pedagogy while I practiced my colors and ABCs.

My mother would go to the specialty stores where teachers picked
up the materials that they needed for the classroom, and she'd stock
up on workbooks filled with lessons. Then she would assign me
homework. Or we'd go to the library and check out copies of *Char-
lotte's Web* and *Gulliver's Travels*. As my reading skills improved, and
I was able to get through the books on my own, she'd quiz me on
what I'd read. Could I summarize the story? What were the messages of the book? Who was my favorite character—and why? When
our neighbors' arguments filtered through our walls, or they turned
their radios up too loud, my mother would march next door and tell
them to turn down the volume because her daughter was trying to
study.

By the time I started kindergarten, I was so well prepared by
my mother's homegrown curriculum that my teacher would give me
books second graders were reading or send me to sit in a first-grade
classroom, where I'd write basic compositions about what I'd done
over the weekend or my favorite memory.

As I got older, my parents continued to supplement my education.
Every day was a new challenge.

"Sunci!" Dad would say, spreading the *New York Times*, a fixture
in our home, on the kitchen table. "Come read this article to me out
loud."

Inevitably there would be words that I had never heard, let alone
read. But Dad was patient. "Sound it out."

"*Po-lem-i-cal*," I'd say, stretching out the syllables.

"What does that mean?" he'd ask. When I shrugged my shoulders,
he'd give me gentle instruction.

"Well, look it up."

I leafed through the *Webster's New World College Dictionary* we kept in our living room so much that it grew stained and dog-eared. Dad dared me to learn every word on its pages, how to say them, how to spell them, and what each meant. Every day when I came home from school, I'd memorize a new word or two, checking them off with a penciled-in "x," and when our family sat down to dinner, I'd proudly recite what I'd learned.

While there were definitely words in the *New York Times* that continued to stump me, before long I grew confident that whatever I didn't understand, I would eventually decipher. I began to feel the power that came with knowledge, even if articles about a potential new freeway or America's actions in Vietnam were beyond what my classmates cared or ever thought about.

My home life also differed from many of my peers in another way: I had to speak Spanish.

When I was growing up in the 1970s, many Latino households emphasized the importance of learning and only speaking English. That dictate grew out of love and pragmatism. Puerto Rican parents wanted their children to fit in, to be accepted. They wanted to give the white world one less excuse to say that their loved ones weren't good enough, not smart enough, not American enough to be given a job or a chance. They wanted to make it more difficult for narrow-minded people to dismiss their offspring as the "other."

My parents, however, had a different point of view. Speaking Spanish was in part a necessity. Nannie Virginia, my mother's mother, didn't speak much English, so everyone spoke the language of Puerto Rico out of respect for her. And, frankly, if you didn't speak Spanish, you really would miss out on a lot of what was going on, from my grandmother's fussing about what various neighbors should be doing with their lives to good old-fashioned family gossip.

But the ability to speak more than one language was gaining in appreciation beyond our home, starting with a critical moment in the year that I was born. The Bilingual Education Act of 1968 officially ushered in a different way of thinking about multilingualism, with lawmakers acknowledging the need to nurture children who spoke little English by providing school instruction in their first language. My mother, ever the activist as well as an educator, was a champion of such thinking. But beyond the politics of it all, she and Nannie Virginia just felt it was important that I be fluent in Spanish so that I could stay rooted in our family's culture.

So my first words included *"abuela"* as well as "Nannie." I learned to slip between English and Spanish seamlessly, and I grew up believing that being able to say things in two languages was not only a talent, it was a gift.

My education went beyond Spanish and spelling too. Nearly every weekend, my parents and I would hop on the subway, or into our Monte Carlo, and immerse ourselves in the art and experiences of New York City. We would wander through a Matisse exhibit at the Metropolitan Museum of Art, study the names of exotic flora at the Botanical Garden in the Bronx, or gaze at centuries-old texts displayed in the majestic New York Public Library on Fifth Avenue. Broadway plays were also on the menu. I remember seeing *Equus* and *Fiddler on the Roof* at a very young age.

Many of the institutions we visited were free or asked for a small donation. But even if there was a cost, my parents scrimped and saved to fill my life with possibility. Dad was always impressed when his colleagues talked about vacations they took overseas, and though we couldn't afford such trips, he wanted me to be prepared for the time he knew would surely come when I would travel the world.

We'd go to Japanese restaurants, where I learned to use chopsticks. And wherever we ate, my parents let me order first so that I could get whatever I wanted—I especially loved lobster. Then my mother and father would scan the menu for an entrée they could afford with the few dollars they had left. It mattered more to them that I grew accustomed to being in all manner of spaces, that I felt equal with any- and everyone else perusing an art gallery or sitting and dining on prime rib in an upscale steak house, than that they got to eat whatever they most had a taste for. They wanted me to know that I deserved to be wherever I chose to wander.

* * *

Compared to the intellectual vigor of my family life, school often felt rote and dull.

When other kids were staring at the blackboard, I'd be staring out the window. Or I'd doodle, dotting the "i" in my first and last names with supersize hearts and sketching cartoon versions of the boy with the curly hair who sat in the seat beside me.

I also grew mischievous. I'd tear a piece of paper from my binder, use my saliva as glue, and toss a spitball at a classmate's head. I'd giggle and whisper jokes, distracting the other kids. Or I'd raise my hand to make endless requests to go to the bathroom, where I'd dawdle and stare in the mirror. After a pit stop at the water fountain in the hallway, I'd slowly amble back to class.

All that, however, was before I met Ms. Lopez.

I walked into her third-grade classroom at St. Anselm on the first day of school in 1976, took a seat in the middle row, and stole a peek at the petite, auburn-haired woman who was practically dwarfed by her massive desk. Little did I know that she was about to play a pivotal role in reshaping my future.

Ms. Lopez believed in engaging with her students. Instead of reciting a lesson, writing words and figures on the chalkboard, then asking us to jot down what we remembered in silence, she asked lots of questions, challenging us to think on our feet and to be articulate.

My previous teachers had grown tired of my hand shooting up in the air.

"What's the verb in this sentence?" Miss Jackson, my second-grade teacher, would ask.

"That's an easy one," I'd think as I practically waved my fingers to get her to call on me.

"Asunción," she'd say with a weary sigh, "you raised your hand already. Give someone else a chance."

Ms. Lopez had a different approach. Instead of limiting me to one or two questions per day, she would call on me often. One afternoon, when the final bell had rung, she asked me to stay behind.

Being asked to stick around after school wasn't an uncommon request for me. In first and second grade, I'd often be held back to receive a finger-wagging warning that if I didn't stop whispering, fidgeting, and daydreaming in class, my teacher was going to have to call my parents. But I was mystified as to why Ms. Lopez wanted to talk. I actually found her interesting and was much more attentive in her class than I had been in others. What had I done to be in trouble?

Ms. Lopez set aside the papers she'd been shuffling and posed a question no teacher had ever asked me. "You keep raising your hand, even after I've stopped calling on you," she said. "Do you know all the answers, or do you just like to talk?"

"Both," I piped up, wanting more than I usually did to get the answer right.

Ms. Lopez smiled and told me that she would see me in the morning. I was relieved but puzzled. What was that all about?

A few days later, Ms. Lopez came and stood by my desk. My classmates were reading quietly, but she asked me to set aside the book and to concentrate instead on the thin pamphlet that she handed me. It was different from the quizzes and tests I'd taken before, the pages filled with multiple answers that I had to sift through. I pulled out my pencil and carefully filled in the tiny circles.

It went on like that for several days, taking these unusual tests, each a little harder than the one I'd taken before, while the other kids pored over readers or basic arithmetic. Finally, the pamphlets came to an end.

It turned out that I'd been taking standardized exams mandated by the state to be given at the end of the school year to assess what students in each grade, from primary through high school, had learned and how prepared they were for the courses that would come next.

Based on my performance, I was ready to graduate from the twelfth grade. Of course, that wasn't happening. But Ms. Lopez said something had to be done. After consulting with the principal, she asked my parents to come to the school for a meeting.

"Asunción can't stay in the third grade," she said, getting straight to the point, "and she doesn't belong in the fourth. So we're going to recommend that she go to the fifth grade."

That was it. In one fell swoop, I was a fifth grader.

In many ways, that wasn't a good idea. I was tossed from a room of pigtailed and crew-cut youngsters who jumped rope at recess into a sea of kids on the cusp of puberty. From then on, through much of my academic career, I would lag behind my classmates socially and sometimes emotionally, if not academically. I was twelve years old when I started high school and sixteen years old when I went away to college. When I looked at my own sixteen-year-old son, so much like me, and contemplate him having to socialize and fit in with nineteen- and twenty-year-olds, I shudder.

But while all the social implications may not have been thought through, I know that Ms. Lopez made her recommendation because she didn't want me to be stifled. Despite the pitfalls that lay ahead, Ms. Lopez indelibly altered my life in a good way, putting me on a track to strive and excel with a fervor and sense of self that carries me to this day.

* * *

The other turning point in my educational journey was the eventual move to a limestone mansion set on a quiet street on the Upper East Side. When I got skipped a grade, my parents wanted me to go to a different school. And my uncle Ed's stabbing had lent their search a new urgency. Believing Manhattan promised a better quality of life, my parents would head there whenever Dad had a day off to scout apartments. Soon, though, they noticed a disturbing pattern.

After checking the *New York Times* or the *Daily News,* it seemed that apartments listed as available when my parents set up an appointment over the phone were no longer available once they showed up in person. They started to suspect that this was a racial thing.

Mom, fair-skinned with chestnut-colored hair, started searching by herself. When my aunt Carmen landed an apartment in Stuyvesant Town, a development not far from where Nannie Virginia lived, she suggested that Mom put in an application.

Aunt Carmen's last name was Ramos, but with her white skin, straight golden-brown hair, and green eyes, and Vietnam Vet, white-looking Puerto Rican husband, they found an apartment very quickly. Mom had a similar complexion, but with an African American husband and biracial child, her chances were slim. So before Mom put in her application, she and Aunt Carmen hatched a plan.

When Mom went to the rental office to fill out the paperwork, she wrote that she was married, but she said that her husband's name was simply William, instead of Willie Moses. And though Mom still typically used her maiden name, on the application she chucked the clearly Latin-sounding Rosa Beza for Rose Cummings.

In a way, Mom was engaging in a version of the subterfuge known as passing, a way of being that black and Latino people pale enough to "pass" for white had practiced for generations.

Passing could be situational, used in an emergency, such as when there was an urgent need to use a "Whites Only" restroom or to avoid being brutalized or worse for walking through a "sundown town" in the south after dark. It could be opportunistic. Adam Clayton Powell Jr., the legendary Harlem minister and lawmaker, reportedly passed for white for a time while attending Colgate University in Hamilton, New York, to avoid the social isolation imposed on his darker-skinned classmates. Still, others made a more permanent decision to use their European appearance to disappear completely and permanently into an all-white world.

For blacks like my dad, whose appearance wouldn't allow them to pass even if they'd wanted to, there were other ways to try to stave off racism, at least for a little while. Studies have shown that employers are more likely to reject applications with a name that they believe sounds African American, not giving the job seeker a courtesy phone call, let alone a chance to interview. Dad didn't need any scholarly research to figure out that reality.

So even though his proper name was Willie Moses, and all his close friends called him Moe, as soon as Dad started working, he began using the name Bill Cummings. He wasn't denying who he was or ashamed of the name that his mother and father had given him. Dad was simply doing what was necessary to get his foot in the door before it could be reflexively and unjustly slammed in his face.

Similarly, Mom was doing what was necessary to get her family a more comfortable life away from the South Bronx. A couple of weeks after she submitted her application, Mom got the good news: the apartment was ours.

The day that we moved in was the first time Dad and I saw our new home. It was also the first time that the man who was there to give us the keys got a look at our multihued family. He looked like he'd seen a black-and-tan ghost.

"Umm," he said, nervously clearing his throat. "I'm expecting Rose Cummings."

"Yes," Mom said with a bright smile. "I'm Rose Cummings."

And, he said, his voice choking as if he feared the answer he knew he was about to hear, "William Cummings?"

"That," my dad said with his deep baritone, "would be me."

The rental agent fumbled with the keys, opened the door, and walked away with a befuddled look on his face. When he'd gone and we three were alone, my parents burst out laughing. I laughed too, though I wasn't sure why. My parents' happiness was just contagious.

Once again, we were on the top floor, the twelfth, this time in Apartment 12A. But that was about the only similarity our new home had to the St. Mary's Houses.

\* \* \*

Stuyvesant Town was an eighty-acre oasis in Manhattan's East Village. The views were spectacular, the jeweled spires of Manhattan's skyline visible from one set of windows, and acres of trees made Stuy Town feel like living in Central Park. There were tennis courts, a huge playground, and gushing water fountains. Unlike St. Mary's, the neighbors didn't have to volunteer for patrol. There were security officers keeping watch from actual guardhouses, making you feel safe playing outside long after dark.

It all took a little time to get used to. When I first saw how the kids left their bikes outside, I thought, "How can they do that? Isn't someone going to take them?" But when morning came, I saw that each and every one was still there, and soon enough my own Schwinn ten-speed was lined up right beside them.

Just like in the South Bronx, I remained cocooned by family. Our apartment was at Fourteenth Street and First Avenue, and Aunt Carmen, along with her husband, my uncle Joey, and their son, Jeffrey, lived only a block and a half away. My aunt Inez lived two blocks away on Twelfth Street. Meanwhile, there was less than a mile between our homes and Nannie Virginia's on East Third Street and First Avenue. My cousin Magaly and my grandmother's brother Emilio also lived in the building. My other cousin Tamara; her husband, Kenny; and their son, Koching, lived a few short blocks away.

Mom, Dad, and I had a two-bedroom apartment with gleaming hardwood floors. My bedroom had a huge closet and a princess bed that my mother covered in peach and white quilts. We had a dishwasher, a new refrigerator, and an actual dining room for the first time.

The worry that so often lined my mother's face before now rarely appeared. She blossomed being around the corner from her sister and so close to her mother. And Jeffrey, an only child like me, was more like a brother than a cousin. He was one year older, but since I'd been skipped ahead, we wound up in the same grade at Immaculate Conception, the local Catholic school directly across the street from our new apartment.

Immaculate Conception went only through the eighth grade, so within a couple of years, it was time to start thinking about where I would go to high school. We lived across the street from Stuyvesant High, arguably the best public high school in New York City, if not the country. I saw the kids, lugging their backpacks, streaming in and out of the wide front entrance, and gathering in clusters after school, and I dreamed of walking among them. After I took the required entrance exam, I was confident that I'd done well and I figured that I was on my way.

But when Dad took a tour of the school, he didn't like what he saw. It was the dawn of punk rock and New Wave music, when Depeche Mode and the Sex Pistols were topping the charts. Kids sported hair streaked purple and neon blue and had piercings seemingly everywhere they could find a soft patch of skin to puncture. They were just being young, expressing themselves as they attempted to figure out who they were, but Dad didn't appreciate their teen angst and creativity. He thought the kids were just plain weird. "Social renegades," he called them.

We also visited Bronx Science and Brooklyn Tech, two other local high schools renowned for their academic rigor. But I continued to think Stuyvesant was perfect—top ranked and close by.

I thought Dad would eventually see the light. But if you'd jotted the names of those three schools in a multiple-choice quiz and presented it to Dad, he would have checked the box "None of the above." He was still searching.

By then, my father had taken a job with Colgate-Palmolive. One day, when he was talking about our visits to various schools, a colleague suggested that he check out a prominent girls' academy on East Sixty-Eighth Street.

Dad took a drive past it one afternoon, and without even going inside he gazed at the graceful facade and made a decision. That was the kind of place he wanted for me. He and Mom arranged for me to take the mandatory entrance exam. In the fall of 1981, I was bound for Dominican Academy.

*  *  *

Dominican Academy was an all-girl Catholic finishing school housed inside a mansion between Park and Madison Avenues, the most exclusive zip code in New York City. Run by the order of the Dominican Sisters, it was founded in 1897 and had a reputation as one of the finest Catholic girls' high schools in the United States.

The first time I walked into the building, primped and pressed in my starched navy-and-white uniform, I thought that I had never been anywhere so grand or seen a space so regal. The front hall was lined with marble, and a deep red carpet muffled our footsteps as we walked up and down the massive staircase that rose in the center of the hall.

To enter Dominican Academy was to be transported to another era, enveloped in a world of age-old manners and tradition. The library where I did my homework was fragrant with smoke wafting from a wood-burning fireplace. There were father-daughter dances at Tavern on the Green, where we waltzed beneath shimmering chandeliers. And when our class of fifty graduated, we wore white lace gowns and carried bouquets, like debutantes coming out at an elegant cotillion ball.

Dominican was also a place of rigorous scholarship. I studied Latin for four years, along with English literature and science. For our classes in art history, my classmates and I took trips to the Frick Collection, an institution that is still one of my favorite places in the whole world. I loved Dominican from the start. Going there made me feel lucky and special.

Yet despite all the grandeur and decorum, my mischievous streak hadn't disappeared. My trek to school involved two subway trains, and I was often late. Since I'd get the same demerit and detention if I were an hour late as I would if I missed the homeroom bell by a minute, many days I decided to treat myself to breakfast at Magnolia's, a bustling diner down the street from school on Madison Avenue.

One morning, when I spotted some friends who were also late, I told them about my late-bell detour and encouraged them to come with me.

We were having a ball, sipping freshly squeezed orange juice and eating pancakes when we should have been immersed in first-period algebra, when suddenly, looming above us, was Sister Timothy, our school's principal.

I have no idea how she knew where to find us. Maybe there were so many of us missing from class Sister Timothy figured something was amiss and decided to comb the Upper East Side in search of her wayward students. Anyway, I didn't spend much time trying to figure out how we'd been discovered. Sister Timothy sometimes seemed as omniscient as the Lord we revered. She knew just by looking at a bowed head or a sluggish walk that you'd forgotten your homework or slipped up on a test. You couldn't get anything past her, and I knew because I often tried.

Hovering like a shadow in her black-and-white habit, which covered her entire head but for her face, Sister Timothy asked who the ringleader was. It wasn't exactly an "I am Spartacus" moment, with everyone rushing to take the blame, but my girlfriends didn't immediately rat me out. I spared them the trouble and fessed up.

Sister Timothy told my schoolmates to settle the bill and get back to school, but she asked me to wait.

"I appreciate your taking responsibility," she said, her lips tightly pursed, "but you're going to have detention."

At Dominican Academy, even detention had the patina of purpose. My after-school punishment was to polish the silverware until it gleamed. I would get much practice figuring out how much wax was necessary to turn the tea service from a pale copper back to a steely gray, since my detention lasted weeks. But no day was as awful as the first. When I was done and headed to Sister Timothy's office to say goodbye, I found my mother and father sitting there, waiting. I felt ashamed. You weren't supposed to embarrass your family, especially not when they were working overtime and doing without to come up with the tuition to send their child to a private school off Park Avenue.

* * *

There was only a handful of students of color at Dominican, and I was one of just two black girls, yet I was always welcomed. Still, perhaps inevitably, there were times when I didn't feel completely comfortable.

During spring break, aside from an occasional trip to Puerto Rico, I usually spent the week at home in my neighborhood. Maybe I'd go to Philadelphia to see my cousin Jeffrey, a high school track star by then, run in the prestigious Penn Relays. But I'd be back in the car and home to Manhattan a few hours later. None of that was even worth a mention when we'd get back to school and the other girls chattered about their family ski trips to Colorado, flights to Europe, or days spent at vacation homes on Long Island.

For the first time in my life, I became self-conscious about my working-class roots.

The South Bronx was still in my heart. But for a long time at Dominican I refused to claim it. When people asked me where I was from, I was quick to blurt out Stuyvesant Town. I didn't want the kids there to know about my Boogie Down world. Not yet. Maybe not ever.

Similarly, when I went home to the Bronx, I didn't speak of the etiquette lessons that I took, the library with the wood-burning fireplace and green velvet walls where I studied, or the magnificent staircase swathed in red carpet that I walked each day. If I did, I figured I'd be branded a traitor for real. And perhaps, for the first time, it would be a title that I deserved. I felt like I was sitting at the center of a Venn diagram, in a bubble all by myself.

I was learning something that the nuns couldn't teach—how to code-switch. Just as my father understood that he would be treated with more respect when he made a call and said that his name was Bill Cummings, and Mom knew that she needed to leave Dad and me behind if she was going to have a chance at securing a decent apartment, I knew instinctively that I had to shift depending on the space I was in.

For African Americans and Latinos trying to circulate and make it in the larger white world, code-switching is a fluidity born of necessity, so subtle that most whites have no idea that we do it, though it requires great skill and can be exhausting.

When I entered Dominican each day, I checked my Bronx-bred lilt at the door. I didn't roll my r's or pepper my speech with the slang or Spanish colloquialisms I used at home. I still loved to dance across our living room to the music of Donna Summer, Shalamar, and Kool and the Gang, but at school, I was listening to Led Zeppelin's "Stairway to Heaven" or the head pounding guitar licks of "You Shook Me All Night Long" by AC/DC.

Looking back, it feels somewhat surreal that having begun my life in tenement walk-ups, I would wind up in a school like Dominican, but I know why my father wanted me to have that experience. It wasn't just the first-class education I received in subjects like history and biology. I was also instructed in the finer points of setting a table: "Start with the salad fork and soup spoon, then work your way in," instructed Sister Christine.

Those are different ends of the learning spectrum, to be sure, but I quickly recognized that knowledge of such things was important. They were the nuances you needed in order to crack the code, the confidence boosters that made you feel comfortable in corridors of power and that allowed you to hold your head high when you encountered people who made assumptions about you based on where you grew up or the color of your skin.

I had what amounted to an entire team working to expand my horizons. In my memory, Sister Timothy and I had a love/hate relationship, but as an adult standing on the foundation that she helped build, I feel nothing but love. She would scold me when my midterm grade in chemistry slipped below an A-, and reprimand me for not bothering to even move my lips during Mass the morning after I'd stayed up late reading *Lord of the Rings*. But I know now, and sort of knew then, that she was hard on me because she knew what I was capable of and she didn't want me to ever sell myself short. She'd spotted my potential just like Ms. Lopez had. That is what great teachers do.

I am where I am in large part because Ms. Lopez and Sister Timothy believed in me, taught me, and pushed me to reach high.

Of course, I was also fortunate to have parents who believed in me and made sure I understood the value of education, building up my confidence and knowledge at home. But the truth is that many parents, just like mine, have hard jobs, long days, and significant responsibilities that they have to tackle at the same time that they are trying to motivate their children. Teachers are the adults who spend the bulk of the day with young people, playing a critical role in shaping how they see the world. The impact that they can have on a young mind is immeasurable.

Unfortunately, too often, they don't have the tools or get the respect that they deserve, and when those charged with educating get short shrift, so do our children, and that fuels the inequality so rampant in our country. It pains me to know that there are others like Ms. Lopez and Sister Timothy, and many who have the potential to be incredible educators, motivators, mentors, whose enthusiasm and effort are doused by low pay and ingratitude. We've seen their frustrations erupt in walkouts that have closed schools and filled streets from Kentucky to Colorado to California.

I know firsthand that education remains the greatest engine of equity in American life, able to lift a child like me from a poor neigh-borhood to the heights of the law and TV news. And yet a good one is becoming increasingly out of reach for too many children, black, brown, Asian, and white. For those who don't have the financial means to attend the top schools or go on trips that expose them to cultures around the world, teachers like Ms. Lopez and Sister Timo-thy, whose perseverance and fortitude can set imaginations ablaze, launching children on their way, make the difference.

CHAPTER THREE

# *NEGRITA* WITH THE GOOD HAIR

When I was inside the halls of Dominican Academy, I felt protected and safe. But the few miles that I had to travel each day to get there filled me with dread.

In the morning, I would take the L train from First Avenue to Union Square. Then I would transfer to the 6, which would drop me off two blocks from school. It was a pretty straightforward trip, less complicated than the one tens of thousands of commuters took every day, traversing the matrix of tunnels that connected New York's five boroughs as they hustled to work or rushed to class.

But many days as I dressed, my stomach trembled with nerves. And as I descended underground, put my token in the turnstile, and waited restlessly on the platform, my heart raced as fast as the train hurtling toward me.

The talk about New York being a tough place is justified. It's charged with a feral energy that flows from all the possibilities that are there for the taking, as well as a harshness honed by endless winters and eight million people living on top of one another like blocks in a game of Jenga. There is excitement, but fear too, and plenty of folks are looking to prey on those they perceive as weaker. The 1980s were particularly dangerous, a time when crime was still high and the city's subways looked like psychedelic canvasses, splattered with graffiti.

I was a target during my daily commute for many reasons. I was a young girl, sandwiched on the subway among strangers, with few options for escape as we trundled underground. I also stood out like a sore thumb in my ridiculous uniform, with my navy-blue blazer, white button-down blouse, plaid skirt, and oxford shoes.

It's a trip that I would never allow my son and daughter, raised in the suburbs, to take on their own today, even though my son is an athlete and over six feet tall, because kids raised in far rougher circumstances would be able to spot him and his more coddled background a mile away. Same for my daughter. Unfortunately, I didn't have a choice.

Once, when I was a twelve-year-old freshman, a disheveled man on a packed train pressed himself against me closely. No matter how I tried to move away, he seemed to always be directly behind me. As the train moved along, I could feel him rubbing himself against me. I finally broke away as the doors opened at the next stop. It was only when I got home at the end of the day that I noticed that my uniform's kilt skirt was stained in the back. I dropped it off at the cleaners and never thought about it again. Years later I realized that he had ejaculated on the back of my skirt. To this day, although I do still take the subway, I don't get on if it is too crowded, and I can't get the memory of that disgusting man rubbing himself against me out of my mind.

Another day a man grabbed my butt. Stunned and frightened, I turned to look at him and found him staring straight back at me with a leer and a smirk. The distance between stops was no farther than a few blocks, but it felt like an eternity. I leaped off the train and sprinted to the next car in the few seconds between the doors opening then shutting again. Once I made it inside, I struggled to breathe, as winded from my fear as I was from the mad dash I'd made to move from one section of the train to another.

But of all the incidents I endured on that daily subway ride, none was as traumatic as the morning that I was jumped by a group of black and Latina girls who took a pair of scissors and chopped off my hair.

* * *

They had tormented me for a long time. They weren't on the subway every day, but the panic I felt wondering when they might next show up was nearly as paralyzing as the anxiety I felt when they actually appeared, grouped together on the train, smacking gum, speaking loudly, and popping off insults like firecrackers.

I would feel the heat rise in my face as they taunted me. "Look at those shoes," one would say, as the other girls cackled around her. "I've got to get me some of those! And where'd that skirt come from?"

Sometimes there was no laughter, just a stream of vicious barbs. "Yeah, you'd better look the other way! You think you're so cute. Yellow bitch."

The day that they jumped me, the attack must have been planned. Some people may have walked around with a small knife in their back pocket or a can of Mace in their purse for protection, but even in New York, I think carrying a pair of scissors was unusual.

I'd felt somewhat relieved when I spotted the clique and for probably the first time the girls didn't yell at me. But in hindsight I should have realized that their silence was ominous.

Typically, when the train reached Sixty-Eighth Street, I would rush ahead to outrun the girls' words and wrath. They never chased me, content it seemed to confine their bullying to our trek along the subway tracks. But this morning, they were on my heels.

I was headed toward the stairs when suddenly they were all around me, a swarm of puffy jackets and heaving book bags. Before I knew what was happening, someone lunged at me, cutting off one of my two pigtails. I crumpled to the ground, crying, my hair strewn beside me. The sound of my tormenters' laughter echoed as they fled toward the exit.

I was hysterical. A couple of other bystanders actually cared enough to ask me if I was all right, but I ignored them, stumbling up the steps toward school. I went straight to the main office, where Sister Timothy called the police and my parents. The officers who arrived a short time later asked if I could describe the girls, and they even escorted me back to the train station to go over what happened. But I didn't have much to tell. It wasn't like I'd ever gotten a good look at the girls to memorize their features, the insignia on their backpacks, or the colors of the clothes they wore. I was intent on avoiding their gaze, not staring right into it.

And while I suspected that they may have attended Julia Richman, a public high school down the block from Dominican, I was street-smart enough to know that if I did identify them, things might get even worse for me.

I kept my thoughts to myself, including when the investigators, Sister Timothy, and even my parents asked me what might have been the girls' motives. Did they try to rob you? Did you get into an argument with one of them one day? Did you bump into them by mistake when boarding the train?

Nope. I knew that the reason they attacked me was lying on the floor of the Sixty-Eighth Street subway station. They looked at my waist-length hair, at my light skin, and hated me for them.

* * *

The trip may have been daunting, but once I got to Dominican, I had various sets of friends. But among the groups I drifted between, there was a clique comprised mostly of girls of color. It was like our otherness was a gravitational pull, drawing us together.

Arminda Aviles was Latina, hovered just under six feet tall, and had dark tresses that dangled at her hips. Margaret Besheer was Lebanese, with a round face and a halo of frizzy hair. Sandra Juanico was Filipino, and wore her dark hair in a blunt cut, with bangs that stopped just short of her piercing eyes. She lived near me, and the two of us would explore Filipino markets together, where I would happily spend my allowance on sampalok (sweet and sour *tamarindo*) and other Filipino foods that I loved so much.

There was also Natalie, a gorgeous Portuguese girl who I now realize was in a pretty scandalous, maybe even illegal relationship. She was sixteen, and her boyfriend was thirty-three. She had her own separate group of friends, satellites who rotated around her glamorous presence. They all seemed so sophisticated with their gloss-stained pouts and jaded observations of the world. I found them fascinating.

Giana "Gigi" Walker was my best friend. She looked like actress Nia Long, with her almond-shaped eyes and beautiful skin. We were inseparable.

But even with Gigi, and my other friends, I stood out, this time because I was so much younger. Some of the students with blue-collar roots like mine got their working papers at sixteen, along with their driving permits. But I wouldn't be sixteen until my senior year. There were also clandestine clubs in Manhattan that welcomed kids that age, but even if I'd lied and pretended to be older, the doorman would likely have laughed in my face. I was barely in my teens and looked it.

So, when my buddies were taking their driving tests, working after-school gigs, or hitting a club on Friday night, I was usually heading home, burying my head in Nancy Drew mysteries or *The Lion, the Witch and the Wardrobe*, jotting down my thoughts in my diary or plumbing my brain to figure out a nine-letter synonym for "dinosaur" as I filled out the *New York Times* crossword puzzle.

Though there weren't many evening activities that I could participate in, I flung myself into pretty much every sport and extracurricular activity Dominican had to offer. I was the yearbook photographer, in the dance club, and a member of the softball team. I also continued to exercise the academic discipline instilled in me by my parents and Ms. Lopez at St. Anselm. My grades, for the most part, were stellar.

Still, it annoyed me that there was always something that set me apart, be it my Bronx roots, my family's working-class status, or my mixed heritage. Being and looking younger than my schoolmates was just one more spoke in that wheel.

Despite being fine with my skipping a grade, which put me in an uncomfortable place socially, my parents still felt it was important that I look my age, so I wasn't allowed to wear makeup. Not a hint of lip gloss or mascara. As a mother myself now, I can relate to their concerns. I was literally a thirteen-year-old girl in high school. But I still felt terribly self-conscious about my appearance.

Once, I asked my parents if I could get a blowout to straighten my naturally wavy hair. My parents didn't think twice.

"A blowout?" each asked me, practically in unison. "You don't need a blowout. Forget it."

But I was tired of corralling my curly mane into ponytails or braids, looking to my eyes like I was a few grades removed from kindergarten. I went to a local pharmacy and bought a lye relaxer. Tucking it in my book bag and practically racing home in the brisk autumn air, I felt rebellious, defiant. My parents weren't home, so I had the apartment to myself. I went into the bathroom and shut the door.

I carefully read the directions, then applied the chemical—the white, cloud-like suds swathing my scalp. But I guess I left it on too long. My head felt like it was on fire, and as I turned on the faucet, dunking my head beneath the stream of water, clumps of my hair began to tumble into the sink. That was the first and last time I ever relaxed my hair.

The remnants of that relaxer disaster can be seen in my senior photo. I look a mess. I had to cut my hair pretty short to save what was still healthy. And since I wasn't allowed to wear makeup, I didn't know how to apply it. Gigi tried to help me with some lip gloss and eyeliner, but compared to the other girls, I looked like a kid trying to play dress-up. I look back at that picture and see all the confusion of what it was, and is, to be me.

* * *

My cousin Jeffrey and his friend Louie escorted me to school for a week or two after I was attacked in the subway, keeping a lookout for the girls. But La Salle Academy was far downtown, so they couldn't do that forever. Pretty soon I was on my own once more.

I tried to figure out safer ways to get there. Sometimes I would take the bus to avoid being trapped underground, but that would take for-ever, crawling the dozens of blocks uptown. On warmer days, I might walk the more than fifty blocks to Dominican, preferring to arrive hot and sweaty than having to brave one more train ride.

That's why I was often late, missing the 8:30 A.M. starting bell. I'd get reamed out by Sister Timothy, of course. But I was willing to take the demerits, the tardy slips, even the occasional round of detention.

What Sister Timothy and even my parents didn't understand was that I was trying to reconcile two worlds: the frenzied, turbulent ter-rain of the city and the cloistered environs of Dominican. I needed to fit in on the way to school and then switch again to blend in once I arrived. Those were just the truths of living in the gray.

* * *

In the hierarchy of isms, race reigns supreme, but it has many off-spring, and one of them is colorism—the hierarchy of complexion that exists among oppressed peoples. Being the child of a pale-skinned Puerto Rican Jewish woman and a copper-toned African American man, I experienced the stepladder of pigment in all its complexity.

My father's people fell on the darker end of the spectrum. With maybe the exception of my great-aunt Doris, nicknamed Red for her rosy undertones, I was the only light-skinned one in the family. And it was apparent from the time I was little that that marked me as different.

My cousins and aunts said that I had "good hair." I remember them being amazed that I didn't run for cover when it started to rain outside—my hair would just get curlier in the rain. They would clamor to play with it, brushing it, braiding it, putting it in updos or ponytails. They'd re-create the styles worn by white actresses on their favorite TV shows; Cindy's ringlets on *The Brady Bunch;* Jaclyn Smith's waves on *Charlie's Angels.* I honestly never understood the fascination.

I stood out at the annual Cummings family reunions down South. Unlike my cousins, my skin would turn red and burn in the Georgia sun. And if my aunts and uncles were describing me for some reason to someone outside the family, the description was usually the same: "You know . . . Moe's girl—the pretty light one."

While some might have reveled in all that attention, it made me uncomfortable because I knew why it was coming my way. All I wanted was to hang out and get along with my cousins. But if they harbored any resentment, they didn't let on. We were a close-knit bunch, jumping double Dutch outside and doing more tasting than helping as our older aunts prepared potato salad, fried chicken, and corn bread.

I was a Cummings, coddled and loved, but Mom most definitely was not. She would put on her game face and make the trip to Georgia, but while Dad and I would have a great time, it was three or four days of misery for her. She was as isolated and left out there as she was in Nannie Mary's apartment back in the South Bronx. It was one of the intricate ironies of race and color that the mixed union that produced my coveted yellow complexion and wavy hair wasn't necessarily welcomed by some of my father's relatives.

It wasn't until June 1967, a mere year before I was born, that the U.S. Supreme Court declared in *Loving v. Virginia* that banning interracial marriage was unconstitutional, allowing relationships that had existed for centuries to finally acquire the privileges and protections of matrimony in every corner of the country. No one in my father's extended family had married someone who wasn't African

American, and while unions between black people and Latinos in New York City were fairly common, we saw few if any interracial couples of any combination when we ventured down South.

My parents' marriage, and my mother in particular, was a frequent topic of conversation among my Cummings relatives. "Umm, umm, umm," one of my great-aunts would cluck. "I can't believe that Moe married that white girl." The other women, sipping iced tea around the kitchen table, would shake their heads in bewildered agreement. Once in a while, someone would say, "Well she ain't really white; she a Puerto Rican."

While a couple of family members would be nice to Mom, most acted as if she wasn't there, leaving her to silently watch us kids play or make awkward attempts to contribute to whatever conversation was going on, only to walk away when she was ignored, barely able to eke out even a cursory response.

It was painful for me to see. It's not easy being the different one, and I understood implicitly that was something my mother and I had in common, though I had it coming from three different directions. I knew what it was like to be the light-skinned girl in a cocoa-colored family, while I also knew how it felt to be the darkest child among my Puerto Rican relatives and to occasionally feel awkward and out of step amid my much older classmates.

What was even sadder in Mom's particular case is that in her consciousness and cultural affection, she was deeply Afrocentric. At the very least, she was the ultimate ally, but more than that, she was immersed in the black American experience and appreciated its powerful history of pushing back on oppression—something that she deeply admired and related to. So it was a particular shame that she of all people would be ostracized by anyone in my father's family.

* * *

While my skin color and long hair set me apart when I was with my father's relatives, I stood out for a different reason when my parents managed to save enough money so that we could visit my mother's extended family in Puerto Rico. There, I had the darkest complexion among my pale-skinned and light-eyed family.

We didn't get to the island very often. It was an expensive trip, and we had to leave behind Nannie Virginia. She'd been orphaned as a child and abused by those entrusted with her care, so Puerto Rico was not a place where she ever wanted to return. But my uncle Joey, married to Aunt Carmen, still had relatives there, and we had many distant cousins, so we would visit when we were able.

Our extended family called me *negrita*, which meant little black girl. It infuriated me.

While that is a common term of endearment in Latin cultures, it never felt loving to me. Instead, I saw it as a label that underscored how my complexion set me apart from my light-haired, light-eyed *primos* and *primas* on the island. I still bristle when I think of it and don't buy it when people say that it is an expression of love.

Like many places in the so-called New World, Puerto Rico was a mélange of cultures, a fusion of the indigenous Taino people, Spanish settlers, and the African diaspora. That mixture made for a multi-colored people whose culture was as deeply influenced by Africa as it was by Europe. But tragically, the same Eurocentric focus found throughout the Americas made an impression on the Puerto Rican people as well, leading them to often prefer their Spanish identity despite the reality that Africa was the foundation of much of the island's music, food, and religion.

Latinos are often in denial about their preference for whiteness, though it's obvious. You'd be hard-pressed to find a deeper-hued Afro Latino or Latina on Spanish language networks like Telemundo, or on the telenovelas that are so popular in Latin America. Dominican hair salons are legendary for the heat and potions they use to blast and smooth every last kink on a customer's head, turning so-called *pelo malo*—"bad hair"—into an abundance of silky, blow-dried strands.

How I was treated depended on the room. In a gathering of Puerto Ricans, if there was a lighter, Penelope Cruz look-alike nearby, I might be overlooked like I wasn't there. But if I was there with my girlfriend Marianela, who was Dominican and darker-complexioned with kinky hair, I would become the magnet. Beauty was defined by the ugliness of colorism.

In Puerto Rico, I was a *morena,* a darker-skinned woman of mixed heritage. I enjoyed visiting the island because it was sun and sand and fun. And I have no doubt my family cared for me. Still, I felt a distance there that I never experienced with Nannie Virginia, my aunts Carmen and Inez, and my other Puerto Rican relatives in New York. Among our family on the island, I was Negrita, Africa obviously running in my veins. And in Latino culture, black was not necessarily beautiful.

\* \* \*

Race is a social construct, a barrier built from an array of stereotypes and assumptions that, however flimsy, have hardened into obstructions used to create a hierarchy of privilege and prejudice in societies around the world, perhaps few more significantly than the United States. It's often assigned based on something as random and illusory as appearance, but the reality is that some darker-complexioned

South Africans have more in common genetically with Norwegians than they do similarly hued people in Ghana.

Whether through voluntary unions or the systemic rape of black and brown women over centuries by white men, there are relatively few Americans whose family trees do not consist of multicolored strands. Still, the "one drop" rule, concocted in the mid-seventeenth century, was yet another shoddy pillar erected to maintain the fiction of white supremacy.

The rule stated that the most negligible amount of African ancestry meant that you were to be considered black, as though it were a stain. Yet a rule hatched in hate became a bond that unified black people. While there have certainly been lighter-skinned African Americans who passed into an all-white world, there have been many others who could have made that choice and decided instead to stay fused to their blackness.

Still, it's not all clasped hands and "Kumbaya" in the broader black community. It's an inescapable truth that black folks, like Latinos, have plenty of hang-ups around color.

To withstand the brutality of slavery, to resist being crushed under the heels of American apartheid, to inveigh against the crippling legacy of bigotry that still infects virtually every institution in this country is to be irrefutably resilient. But it is hard to deal with a near-daily assault of outrages and injustices, microaggressions and outright abuse, without wearing down. All that can take a toll on the psyche, on your self-esteem, on your soul. And in a society relentlessly, and sometimes violently, dominated by those of European descent, it can take vigilance to not believe that whiteness makes you smarter, prettier, better.

Lighter-skinned blacks, the more obvious manifestations of intimate, often nonconsensual relations between white men and black women, were treated marginally better in the racist antebellum South. I believe that I too, on more than a few occasions, have won a competition or garnered a coveted gig because I have lighter skin and straighter hair. That's just the plain truth. And when dealing with other black people, the favoritism I gained because of my appearance was apparent pretty much every day.

While living in Stuy Town, I attracted the attention of all kinds of boys, from Eddie Velardi, an Italian American kid in the neighborhood, to Louie, my cousin Jeffrey's friend, who escorted me to my senior prom. But when I was with an African American crowd, it was like I was Vanessa Williams, Halle Berry, and Rihanna all rolled into one.

On the few occasions when my parents allowed me to go, I was inevitably the belle of the house party, first to be asked to dance and out on the dance floor all night. When I was walking down the street with Tanya, one of my two best friends from kindergarten, who had kinkier hair and dark brown skin, boys we encountered might joke and chat with both of us, but they would be staring at me. Word would inevitably trickle back: "Johnny likes you"; "Derrick thinks you're really pretty"; "Andre says you're the finest girl in the neighborhood."

But, actually, I wasn't. I would look at my friends and know that I wasn't the most beautiful—not by a mile. But I was the lightest. Maybe because they still had boyfriends and dates, my darker-skinned friends seemed to shrug it off, but I figure there had to be moments when they were hurt by the outrageousness of it all.

* * *

But the hurt cut both ways, and depending on the context, I also learned the pain of rejection based on the darkness of my skin. When

I was around thirteen, I had a crush on that Italian American boy named Eddie Velardi. He had thick brown hair styled in a DA. He was so cute, and the cool thing was that he liked me too.

There was a dance happening at a recreation center in the neighborhood, and Eddie asked me to be his date. I was so excited, figuring out how I was going to convince Mom and Dad to let me go and sifting through my closet to figure out what outfit I would wear.

But a few days later Eddie had bad news. He couldn't take me.

He approached me as I was sitting outside with an Irish American kid from the neighborhood, Richie Conroy.

"Hey," he said, nervously looking down at the sidewalk. "Sorry to say this, but we can't go to the dance."

"Why?" I said, whipping my head around.

He paused. "My mom said black cats don't screw birds."

Richie and I stood there confused. What the heck did that mean?

"She asked what you looked like," Eddie said, "and I mentioned that you were black. So she said that I couldn't go out with you."

I couldn't believe what I was hearing. I stood there in shock. Finally Richie broke through the awkward silence.

"Your mother's stupid. Asunción, you want to go with me?"

I didn't miss a beat. "Yeah!" Richie also asked if he could give me a ride on the handlebars of his bike. As we rode away with Eddie in the background, his mouth hanging wide-open, I felt like my heart was beating through my chest. It felt like I was flying. I will always be grateful to Richie for that moment.

My parents, perhaps motivated by my story about Eddie's ignorant mother, let me go to the dance. And I was proud to be with Richie, who happened to be cuter than Eddie anyway. He told me that I was one of the prettiest girls he knew. We had a blast.

Richie's request, so quick to follow Eddie's rejection, made the sting I initially felt fade quickly away.

But it hurt. And I've never forgotten.

* * *

The obsession with complexion was apparent even when I grew up and went on to college at SUNY Binghamton. During my freshman year, I met one of my best friends, Kathy, the daughter of Haitian immigrants, with slate-sharp cheekbones, deep-set eyes, and a Naomi Campbell straight weave. Kathy was drop-dead gorgeous.

Yet it was the same old dynamic. Though white men recognized and raved about Kathy's classic beauty, when we encountered black men, they would always approach me, commenting on how nice I looked that day or asking for my phone number. I felt bewildered by it all, but Kathy was resigned.

"That's just how it is," she would say on the few occasions when I broached the uncomfortable topic.

In 1987, I became the first non-white woman to be named Miss Binghamton. I was attending college on a scholarship and also had work-study jobs, filing books in the library and ringing up orders in the campus food court to make ends meet. My parents and grandparents always sent money, but I hated relying on that. One day on campus, I noticed posters advertising the campus beauty pageant and its $2,500 prize. I decided to enter as a lark. Besides the money, there was also a promised trip to the Bahamas, a place I'd never been. After I won, I was stunned to discover in an article in the local newspaper that every Miss Binghamton before me had been white.

"Can you believe that I'm the first black chick to be Miss Binghamton in the history of this school?" I asked a group of classmates, tossing down the newspaper on a bench as we milled around the common outdoor space that we dubbed "the yard." "How stupid is that? It's 1987."

I expected the group, mostly African American, to agree, echoing my incredulous outrage. But that's not the response I got.

"Of course you won," said my friend Mary, whose deep-brown complexion mirrored Kathy's.

"Yep," another friend, Betty, chimed in. "They know it's about time, and if they're going to give it to anybody not white, it would be you. Your look is acceptable. You're as light as they come."

At first, I was angry. But I thought about it a few seconds, then piped down because I realized that they were right. Our college was in a pocket of New York that, when it came to racial attitudes and diversity, was closer to a rural town in the Rust Belt than it was to my hometown of New York City. In the 1980s, there was a lot of racial tension, from black people being called the n-word by drunk locals on a Saturday night to security officers hassling the few black students on campus. If I had looked like the beautiful Oscar-winning actress Lupita Nyong'o, I probably wouldn't have made it to the final round of the pageant, let alone been named Miss Binghamton.

There would be other instances where I believe my complexion, my "look," would give me an advantage with my white peers, from courtrooms where I argued cases, to perhaps my first jobs in broadcasting. Still, as unjust and troubling as all that is, for many black people, I believe the color-based bias among our own is the most painful. Even organizations that I now belong to and appreciate, like my sorority, Alpha Kappa Alpha Sorority Inc., and Jack and Jill of America, which centers around the nurturing of black children, have historically had a reputation for primarily welcoming those who were lighter. In the past, some black groups and even churches enforced a shameful brown paper bag test to exclude anyone darker than the arbitrary shade of a sack used to tote sandwiches and juice boxes.

In a society where racism is still all too prevalent, such an internalized hierarchy pours salt into a deep wound. It's understandable why some would look at those at the top of that totem pole and seethe with resentment. But what has always been difficult for me to understand, is why there is so little room for people to fully embrace someone like me, someone with more than one cultural identity. For those, like me who live in the gray, or as my friends sometimes say, in the beige, there is so little acceptance. So little space.

# CHAPTER FOUR

# I AM WHAT I AM

One Friday, long after I'd finished school, become a lawyer, and forged a career in journalism, I went out to lunch with a few of my friends and colleagues. I was sitting next to Don Lemon, the CNN anchor I'd been close to since I'd worked at that network as a legal analyst. Across from us was Ana Navarro, another good friend from CNN who'd recently become a contributor to *The View*, where I was now a co-host. And rounding out our circle was Candi Carter, one of *The View*'s executive producers.

We were sharing stories and laughs, day drinking wine, cocktails, and then tiny glasses of limoncello at Cafe Fiorello, a lovely Italian restaurant near Lincoln Center where we were invited to sit in a secluded corner booth kept on reserve for one of my other *View* co-hosts, Joy Behar. After two or three shots, I guess tongues were feeling looser, and the confessions, like the Italian liqueur, began to flow.

"You know, I've got to admit when I first heard you identify as Latina, I just didn't understand it," said Ana, setting down her glass and looking at me in the straightforward way that has made her a favorite with millions of viewers who appreciate her bluntly honest takes on Donald Trump and the state of his and Ana's political party, the GOP.

"Really?" I asked, taken aback. "Why not?"

"I don't know." She shrugged. "You were always so vocal about issues facing the black community, I just never saw you as anything else."

Ana and I had known each for nearly a decade, but she'd apparently spent years thinking I was a pretender. Her judgment didn't start to shift until one day she overheard me speaking on the phone in Spanish to my grandmother. "Holy shit," she told me she uttered to herself. "She really is Spanish!"

Overhearing me speak in Spanish helped prove my legitimacy as a Latina. But J.Lo's mother sealed the deal.

Guadalupe Lopez, the mother of the singer and actress Jennifer Lopez, is a *View* superfan, popping up in the audience at least once or twice a month. We frequently give her a shout-out from the set, and at one taping, shortly before Christmas, I spotted her sitting in her usual front-row seat.

"You've got to get home tonight and start working on those *pasteles*, right?" I asked her in Spanish, referring to the tamale-like combination of green banana and plantain that is a centerpiece of every Puerto Rican Christmas meal. I knew that I was going to be staying up late myself, filling banana leaves with the savory mixture so that I could boil them in hot water on Christmas Eve, then serve them alongside *arroz con gandules* and *pernil,* the slow-cooked roasted pork that is another Puerto Rican holiday staple.

"That's right," she replied, nodding with a knowing smile. Apparently, after witnessing that exchange, Ana knew.

"When I got to know you better, I felt guilty," Ana told me that day at lunch. "I had been so dismissive. And then when I found out your real name was Asunción, I was so shocked, and I realized I'd missed you completely. I have to apologize to you."

Now it was Candi's turn. "I guess I owe you an apology too," she said.

"Oh Lord," I thought, my buzz from the lemony liqueur slowly dying. I immediately knew what she was apologizing for.

Candi, a former senior producer for Oprah Winfrey, was a striking dark-skinned black woman. She and I belonged to the same black sorority, Alpha Kappa Alpha, which I had learned one day not long after I'd joined *The View* and we started chatting about our college days. But despite that bond, Candi had let me know a year earlier that she didn't consider me to be African American.

I'd begun to notice that when the show wanted to tackle issues of particular importance to the black community, such as police brutality or the controversy that cropped up the year not a single African American—or any other person of color for that matter—was nominated for a major Academy Award, I was never consulted about the contours of the conversation, let alone the possible guests.

It made no sense and really irritated me. Ilyasah Shabazz, one of Malcolm X's six daughters, was a dear friend of mine, as was Stephanie Rawlings-Blake, the former mayor of Baltimore and Democratic National Committee secretary. I also knew many of the members of the Congressional Black Caucus and could reach out to them. I couldn't understand why, if there was a show celebrating Martin Luther King Jr. or a segment touching on voting rights, the producers wouldn't approach me about my contacts. I could hook them up with powerful voices and pertinent points of view. So I took my complaint to Candi.

"Well," she replied curtly, "you're not really black."

If I was annoyed before, now I was angry. "Really?" I said, trying to keep my voice steady. "I'm not? Do you think white people don't think I'm black? You think if the KKK comes to round people up, they would bypass me?"

Flustered and needing to get to an appointment, I hadn't had time to dive more deeply into a debate. I rushed off, and in the months that followed, Candi and I both moved on, though our unresolved conversation opened an uncomfortable void between us.

Now, at Fiorello's, Candi acknowledged the mistake she'd made and the hurt she'd seen on my face that day. "I know that my words affected you, and I meant to come back and apologize, but I realize I never did," Candi said. "So I want to say now that I'm sorry."

I sat there, the slight giddiness I'd felt earlier having completely faded away. The conversation Ana, Candi, and I were having in many ways summed up my entire life. Two women, one Latina and one black, had both previously dismissed me. Like so many people, they couldn't understand how I could be of both worlds. How could I be so authentically Black and so authentically Latina at the same time?

Perhaps Ana saw my blackness when at CNN, I vigorously fought for Trayvon Martin, the black teenager murdered by a white vigilante, to be portrayed in all his humanity and for the network to keep a spotlight on the seemingly unyielding string of shootings of other unarmed black men and boys that followed his death. But I was equally outspoken at ABC about Donald Trump's bigoted declaration that Mexicans were rapists, his nativist crusade to build a southern border wall, and his callous indifference to fellow Americans in Puerto Rico whose lives were upended by Hurricane Maria.

Meanwhile, Candi's view of me might have been clouded by the colorism that unfortunately still infects the black community. Perhaps she believed that because my skin was lighter and my hair straighter, my life had to be easier than the lives of my darker-skinned peers. I couldn't dismiss her experience, and I also couldn't entirely disagree. I do believe that my appearance has probably made the road I walk a little less thorny than the one traveled by darker-skinned African Americans.

Still, when I encounter such sentiments from blacks and Latinos alike, even from friends, I feel odd. I believe people with mixed ancestry are the embodiment of the American dream, reflecting the tapestry that has enriched this country. My mingled bloodlines should make boundaries and borders disappear. But too often my very existence makes people uncomfortable.

I sincerely appreciated Ana's and Candi's apologies, and I told them so. But I felt a familiar weariness that came from constantly being placed in a box and questioned when I dared to exist beyond it. It was a lonely place to be.

Don turned to look at me. He and I worked together at CNN for seven years, and he'd known me longer than anyone else at the table. A man who lives with his own dual identities as an African American who also happens to be gay, Don has always been very accepting of who I am and seems to understand deep in his marrow the resistance I often encounter.

He filled the awkward silence. "Welcome to Sunny's world."

* * *

The distance between my two grandmothers, one a Puerto Rican woman living in Manhattan, the other a Georgia-reared black woman residing in the South Bronx, could be covered with a forty-minute subway ride. Though the number 6 train shuttled me between two distinct spaces, I didn't have to change a thing about myself along the way. Nannie Mary, Nannie Virginia, and my relatives loved me for all that I was. But that was not the case as I made my way in the wider society.

Strangers and even new acquaintances wanted me to put my varied identities on a table, then pick one—like I was plucking from a deck of cards or participating in a street game of three-card monte. They often questioned my background because in their eyes, I did not resemble what a Puerto Rican or a black American woman was supposed to look like. I was too brown to be one, and too light to be the other.

What do you do when you love all those who made you who you are, but the outside world doesn't do the same? What do you do when your authentic self is dissected and debated by others who want to deny the dualities that have always made up this country?

Those questions have been a never-ending part of my journey. In a country that still too often makes snap judgments based on race, people want to know what you are in order to decide *who* you are and how they should treat you. But as the United States grows increasingly diverse, moving toward a time when no single ethnic group is in the majority and interracial couples put their love on display in settings ranging from television ads to the local mall, limiting people to a singular identity is no longer so easy. Though folks do try.

I identify as black and Puerto Rican, or Puerto Rican and black, whichever way those identities tumble from my mouth in any given moment. Afro Latina works as well. To be a mixture of those two things is to me the best of both worlds. Black culture in the US is resonant with achievement and creativity in the face of the greatest of obstacles, always willing this country to be its best even as it's so often shown black people it's very worst.

Meanwhile, the Puerto Rican culture is unique within the American mosaic, at once a part of this country, yet so very different. It's not just that the people on the island speak a different language. The music, from salsa to reggaeton, is also singular, as is the cuisine. And the blackness of the island is more aligned with West Africa than with the cultural mores of Puerto Ricans' black brothers and sisters on the mainland.

I was raised to embrace both sides of my family. To choose one identity would have been like giving the cold shoulder to the other, and I could no more have done that than I could cut off contact with one of my beloved grandmothers, say goodbye to family gatherings in Georgia with collards and ribs, or vow to never again eat *pasteles* on Christmas Eve.

\* \* \*

Today, many folks know me as Sunny, but throughout my childhood, I was called by my proper name, Asunción. I didn't become Sunny until I went to college and some of my new friends assigned me that nickname because it was easier to say. Even my husband, Manny, says Sunny in public since that's how I'm known professionally, but he calls out to Asunción when we are at home.

Like my name, which came from her beloved sister, my Puerto Rican identity was largely nurtured and shaped by my nannie Virginia.

Nannie was born into struggle. By the time she was eight, both of her parents were dead, and she was taken into the home of her mother's sister, where she was sexually abused. When she turned sixteen, she took her infant daughter, Carmen Lydia, and what few possessions she had and joined her brother and sister in New York, eager to leave behind bad memories and to make a better life for herself and the family she would one day have.

My nannie was a beauty, with a mane of jet-black hair that danced around her full hips, turning heads long after she'd become a grandmother. With her caramel complexion and doe-like eyes, she resembled the fictional princess Moana. She had four loves, and with each of her first three came a child—my aunt Carmen; then my mother, Rosa; and finally Aunt Inez. But even though there was often a man around, Nannie depended on no one. Her difficult childhood had taught her better than that.

If fate had turned out another way, the mellifluous voice she used to sing me Puerto Rican lullabies could have taken her to the stage on Broadway or at Carnegie Hall. She was a marvelous storyteller, spinning lush tales complete with the gestures and facial expressions of an actor stalking a Broadway stage. And the demeanor that enabled her to turn the racist Hells Angels gang that met near her home into protectors of her family and the block, along with the quick mind that recorded the due date of every bill with just a glance, could have served her well as a diplomat or college professor.

But Nannie Virginia only made it through the sixth grade, and she could barely speak English. So she poured her energies into her three children, who would all go on to have thriving careers, as well as her two grandchildren over whom she doted. She took jobs others looked down on, cleaning toilets and scrubbing floors. And one of her many gifts was the ability to fix just about anything, from a faulty furnace to a fickle carburetor. She became superintendent of each of the small rent-controlled apartment buildings on the Lower East Side where she spent most of her life, expertly handling whatever went wrong in their dozen or so apartments.

People didn't just go to Nannie Virginia when their heat wasn't working. They sought her help with problems that were much more personal, though no less urgent. Around the neighborhood, Nannie was known as "La Bruja," which technically means the witch, though to Puerto Ricans it did not carry the malevolent leanings that translation implies. Rather, it was bestowed on her because Nannie was at one time a practitioner of Santeria.

Santeria is a religion of the African diaspora, derived from the Yoruba culture in West Africa but blended with the Catholicism the enslaved learned from those who enslaved them. To hold on to their original ways of worship, and to also perhaps disguise them from the slave masters, who would have disapproved, the orishas, or deities, of Santeria were given Catholic counterparts. Chango was the king, a warrior who wielded the power of fire and thunder and whose Catholic peer is St. Barbara. Babalu Aye, pushed into the popular consciousness by Cuban singer Desi Arnaz's infectious song "Babalu," was the orisha of healing. He was most closely associated with St. Lazarus. And there were many more whose powers and traits Nannie Virginia could recite in her sleep.

Nannie placed glasses of water by her front door to wash away bad spirits, and inside her apartment, there was an altar known as an *igbodu*, surrounded by flickering candles. People would come from near and far bearing gifts of food to sit and hear Nannie's visions. She would don the colorful beaded necklaces known as *collares*, which marked her standing as a Santeria initiate and I would peek around the doorjamb of the kitchen or bedroom to watch her meet the worried yet expectant gaze of a visitor eager to hear Nannie's pronouncements about their health, the fate of a distant loved one, or their chances at finding fortune and love. Nannie's best friend, a transgender woman named Sylvia, would often join her, the two of them together presiding over ceremonies and "reading" the destinies of those who'd come, sometimes in desperation, seeking answers. One evening she held what I can only describe in my memory as a séance, and I believe I saw someone levitate.

I also accompanied Nannie when she visited the crowded botanica, a shop that can be found in virtually every Latino neighborhood, where she'd buy the herbs she used to create poultices and balms for healing. I'm sure it all would have seemed odd to my schoolmates at Dominican, or unsettling to the crowd I later met and worked with in the U.S. Attorney's office and in newsrooms at ABC and CNN. But it was an integral part of my childhood, forming a sweet spot in my memory that makes me smile whenever I revisit it.

Nannie, who so many people turned to for a glimpse of their futures, said that I had the gift too, the "third eye" that allowed you to tap into another dimension and feel and see what might be coming. I'm not so sure about that, but I have definitely incorporated much of what she taught me. For instance, I learned that I could go to the botanica, buy a candle bearing the name of a certain orisha, and then pray that a person causing me problems would vanish from my life. Or I could write the troublemaker's name on a piece of paper, fold it within a piece of aluminum foil, and stick it in the freezer.

People often misconstrue this belief system, thinking that it's witchcraft or evil. It isn't. The intentions behind it are to better your life, not to hurt someone else. You're not praying for something bad to befall a person. You're simply trying to get them away from you, to freeze them out—literally. There are probably a few pieces of frozen tinfoil in my refrigerator right now. And I'll tell you something else: It works.

There was a producer at CNN who told me once that I didn't have what it took to be a national news personality. Deflated, I complained to my mother, and she told me to put him on ice. So I did. Not too many months later, he moved out of the news division. I figured I'd never hear from him again, but he had the nerve to text me after I'd landed at *The View* to tell me how happy he was for my success. I didn't bother to reply. But I remembered that knot of frozen aluminum foil and chuckled to myself.

Another remnant of Puerto Rico that Nannie transplanted to the city was raising chickens. They roosted right on her apartment building's roof as well as on the fire escape. Her neighbors, many of whom had come from the Dominican Republic as well as from Puerto Rico, never complained. And I'm sure that even if the building owners minded, Nannie was such a good superintendent, they didn't dare make a fuss.

Nannie and I would head uptown to La Marqueta, a marketplace in Spanish Harlem that was as colorful and frenzied as a West African bazaar, to buy chicks and feed. The workers would smile warmly. "Doña Virginia!" they would yell, eager to greet the beautiful bruja who could channel the ancestors and maybe, just maybe, give them reassurance that one day their dreams would come true. These days I raise my own hens, but I don't turn them into soup to be served on Sunday along with *arroz con gandules*. Back then, though, having chickens go from the fire escape to the kitchen table was just a part of my life, as much a family tradition as celebrating Three Kings Day, spinning albums stamped with the legendary El Gran Combo, and displaying the single-starred flag of Puerto Rico in the living room or dangling it from the fire escape. Yet Puerto Ricans often questioned my identity.

I'll never be mistaken for Jennifer Lopez. And in actuality, plenty of other Puerto Ricans wouldn't be either. Like African Americans, they come in a multitude of shades with a range of hair textures. But when I went to college and showed up at my first meeting of the Hispanic Student Union, many of the other members looked at me like I had three heads.

After the official business was over and we all milled about, I was pelted with questions as though I were on a quiz show called *How Puerto Rican Are You?*

"Do you speak Spanish?" one interrogator asked.

"What kinds of food does your family eat at home?" was the not-so-casual query of another, who arched her eyebrows with suspicion.

"*Sí,*" was my response to the first question, and "I would guess the same things you do," was my answer to the second—though when I was faced with such queries, I often wouldn't bother answering at all.

I resented the probing. I was tired of the doubts. When I complained to my parents, they emphasized that it was other people's problems, that I didn't have to split myself into pieces. But despite their reassurance, I found it stressful for my identities to be put up for debate or to feel that I could choose only one part of me.

When I took standardized tests in school, before I could display my knowledge, I had to check off my ethnicity. Nowadays, when more Americans are proudly proclaiming their mixed heritage, it's accepted that people will check multiple boxes on the census or an exam or choose not to check any boxes at all. But thirty-five years ago, neither was an option.

I would color in two circles, one next to "Hispanic" and another next to the word "black." On the occasions when my teacher noticed, he or she would hand the test back to me and tell me that I couldn't fill out more than one. "Why?" I would ask, used to but never accepting of this dilemma.

"Well," my flustered instructor would say, "you just can't."

They were trying to get me to follow the rules, but I doubt they understood how wrenching that was for me. Sometimes I would refuse, and when the test was submitted, the processing machine would spit the exam back out. My dual identities were literally, officially rejected. What does that say to someone like me?

When there was no way around it and I had to pick a single piece of my heritage, I would go ahead and check "black." In a way, that choice flowed from the same defiance that made me initially check two circles to indicate both of my identities in the first place.

Growing up in the 1970s, I had seen the hysterical scenes of white Bostonians reacting violently, and frankly acting nutty, because their kids were being bused to schools with black children. I knew from conversations in my home, and the books I read by Maya Angelou, Richard Wright, and James Baldwin, that black folks had it the hardest, that we were treated the worst, not just in the United States but all over the world. It infuriated me, and perhaps even more than my father, it truly angered my mom.

So if I had to pick one piece of myself in order to be evaluated on a stupid test, if I was going to identify with only one half of who I was, I chose to cast my lot with the more downtrodden.

When I applied to college, I was faced with the same familiar conundrum. There were particular scholarships designated for African Americans and others focused on Latinos. Some, like the Equal Opportunity Program, gave grants to a broad spectrum of students of color, but it still wanted you to specify your ethnic background.

Dealing with such questions continued to rattle my nerves, and when my ability to pay for college was on the line, I was even more torn and confused. What should I do? Which box should I check?

I asked my parents. And Dad didn't stutter.

"Go for everything," Dad said. "Check off as many damn boxes as you can. You're black. You're Puerto Rican. So you qualify for all of that. And besides, you're so smart, the school should pay you to go!"

I did what Dad said and applied for every potential funnel of financial aid that I could find. My dream school was Cornell, which had hospitality and agriculture programs. Channeling the farm-to-table craze about twenty years too early, I thought a bed-and-breakfast that served locally grown food could be a hit. New York University and Syracuse University were also top contenders. Those schools were not really in our budget. All in all, I was accepted to over twenty top colleges and universities. Ultimately, I decided to attend SUNY Binghamton, one of the sixty-four campuses in New York's state university system, which awarded me a full scholarship. At age sixteen, I headed to college.

* * *

After years of having to go straight home from school unless I was hanging out with my cousin Jeffrey, being on my own, in college, was exhilarating. I fell in with a wonderfully eclectic group of friends. Our buoyant bunch would go to concerts together or gather in someone's room to laugh, dance, and groove to the reggae jams of Third World and Erasure. It was as magical and fun as you would want college to be.

Away from the watchful eyes of my parents, I also got my first boyfriend. Eric was a year older, tall, a mixture of black and Portuguese and, I thought, fine as hell. He wasn't faithful and in reality was untruthful and manipulative. But we also shared a lot of good times and would date off and on until I graduated.

Eric introduced me to black Greek life. He was a member of Alpha Phi Alpha, the first black fraternity, which was founded at Cornell University at the start of the twentieth century. Black fraternities and sororities had some similarities to their white counterparts with their pledging processes and secret rituals, but like so many parts of American culture, black folks infused Greek life with their own flavor.

One of their most glorious traditions was "stepping," the syncopated dance form that was spotlighted in films like *Stomp the Yard* and on the television show *A Different World,* which chronicled life on a historically black campus. At step shows, chapters would compete, executing intricate dance routines in unison, sporting their colors, calling out their letters, and yelling chants in what would become a cacophony of dueling choruses. Members would travel from all over the region or country to attend, and it would be a raucous, joyful, exuberant display of blackness.

Even more than the shows and parties, I was impressed with the community service demonstrated by the Alpha Kappa Alpha sorority members at SUNY Binghamton. I eventually pledged and became part of its sisterhood. After living between worlds for my entire life, it felt particularly poignant to find myself fitting squarely in a scene that felt real and *right*. I went from straddling worlds to quite literally stepping defiantly, deliberately into a space I could bring my whole self to.

* * *

Later, when I became a host on *The View*, I'd have to tap back into my reserves to maintain that deep connection to my blended roots. I became used to people getting angry at me when I spoke Spanish on air or discussed the legacy of Jim Crow one minute and the argument for Puerto Rican hurricane aid in the next. I was bombarded by critical emails and tweets. Once again, I was being told to pick a single lane, and stay in it.

*"Why Do You Speak Spanish? You are Black. Stop trying to act like you're not!"*

*"Why are you black one minute, and Latina the next? You need to quit."*

But I would no longer hear that from my friend Ana Navarro. After she began appearing on *The View* every Friday, when Whoopi Goldberg was off, our friendship grew.

Her dressing room is a couple of doors away from mine, and when we meet up in the hallway, go out for lunch, or just hang out, we often banter in Spanish. It's a tie that binds us together, a taste of home that we share in the office.

Ana also knows about my hens, over a dozen that I tend to each morning. I find it therapeutic before I head to the city and get pulled into the cyclone of my day to check for eggs, scatter kernels of feed, and hear the quiet clucking of my birds. Those tasks also make me feel close to Nannie Virginia, who has passed on, and to the island where she was born.

"You are so Puerto Rican!" Ana said one day after I'd gone on and on about my hens on *The View*.

Another time, I was speaking to an African American friend of mine about how I had to pick up some collard greens so that I could clean them and have them ready to cook on Thanksgiving Day.

"Man," she said, shaking her head and smiling, "you are so black!"

My reply to both? Yes. And yes. I am.

# CHAPTER FIVE

# IN THE SYSTEM BUT NOT OF IT

In my life, I've had many jobs and ambitions—radio host, waitress, journalist, lawyer. For a time, I even wanted to be a nun.

Though my grandmother practiced Santeria, like many immersed in that tradition, she was also fiercely Catholic. She went to church several days a week and wore a cross around her neck, layering it with the *collares* she donned for Santeria ceremonies. My mother was also religious, taking me to Mass most Sundays, though the same eclectic spirit that led her to embrace black American culture also inspired her to honor her Jewish father's heritage by hosting the occasional seder.

In many ways, when it came to our faith, we were old-school. Nannie and my aunts believed that a parochial education was best, perhaps because the religious rites mixed in with reading, writing, and arithmetic made those institutions seem more disciplined. And though we did not share the same blood, we considered the leaders in our local church to be like family.

Father Francis Burns was the priest at Nativity, Nannie's church on the Lower East Side, and he would often come over for dinner. If Nannie had owned a red carpet, she likely would have rolled it out whenever he came to call. The table would be laden with *arroz con gandules* or *pernil* or *pollo guisado*. Father Francis was Irish American, with silver hair, blue eyes, and an appetite even larger than his hefty frame. He had a bit of a temper, but an even stronger sense of humor, and he would regale us with stories as he ate and sipped a glass of red wine that Nannie never let go empty.

We were also very close to Father Francis's sister Eileen, who was a nun. Like many of the sisters who taught me at Dominican, she was pious and brilliant. Many people think of a nun and picture someone who is rigid and unrelatable. But Sister Eileen was warm and funny, the type of person you knew you could approach with your most heartfelt questions and worries. I would look at Sister Eileen and think that was the type of woman I wanted to grow up to be.

In my senior year of high school, I attended what was for students at Dominican a mandatory religious retreat. It was called Veritas, which is Latin for "truth." The overnight trip at a convent in upstate New York immersed us in ritual. Each of us was given a small wooden cross, and we would pray in solitude. Paradoxically, we were able to see the Dominican Sisters in a more relaxed mode. It was the first and only time that I ever saw my English teacher, Sister Christine, in casual, everyday clothes. Away from the all-seeing eyes of Sister Timothy, she and our other teachers joked and laughed freely. It was a joy to see these women, so close to God, behave in ways that were so natural and down-to-earth.

There was a music night, during which Sister Christine performed with a few other nuns to Aretha Franklin's "A Natural Woman." We all sat around and cheered. It was one of the best times I had in high school.

Growing up with an activist mother and a grandmother who was a matriarch for the entire neighborhood, I've always felt it was important to help others. And what higher calling could there be than to provide solace and sustenance to a community, to a religious flock, by being a nun? After that weekend spent seeking truth, I thought that maybe my truth was to become a part of that sacred sisterhood. I told Mom my potential plan when I got home.

"I really enjoyed the retreat!" I said. "I'm becoming a nun."

My mother could have come back with a couple of different responses. She might have yelled, "Hell no!" But that would have been particularly inappropriate given the subject. Mom decided instead to walk me back from the edge of the cliff gently.

"You can help people in a lot of ways," she said, continuing to fold laundry. "You could go to medical school or volunteer at a shelter. You don't have to spend your life in a convent."

In hindsight, Mom probably wanted to preserve her chance at one day having grandchildren. Whatever her motivation, she needn't have worried. My desire to become a nun vanished rather quickly as I became consumed by college applications, senior activities, and weekends spent hanging out with my friends.

When I went off to SUNY Binghamton, however, and began studying communications, a new ambition began to burn inside me. And that dream wouldn't be doused so fast.

The C I got on my first exam in a science course during my freshman year made me quickly realize that my road to medical school could likely be littered with mediocre grades, so I needed to find another major. Based on how much I enjoyed writing as well as keeping up with the workings of government and current events, it was a toss-up between political science and journalism. I ultimately settled on the latter because I wanted to hone my ability to tell stories. And as I learned to edit and weave narratives with video, I zeroed in on a possible career in broadcasting.

But Mom remained fixated on my becoming a lawyer. It would come up constantly: when I was home from school or when we took a few minutes during the week to catch up on the phone. I think that had always been her personal dream. At one point, she even took the LSAT, the exam required to enroll in law school, though she never pursued a law degree. I listened to her not-so-subtle suggestions but didn't intend to act on them. I envisioned myself having a more creative career.

When I graduated from SUNY, I, like several of my classmates, decided to take some time off, to figure out exactly what was next. My dad was cool with that idea. "What are you, twenty, twenty-one?" he said. "You should travel. Have fun. You've got time."

Mom was less patient. She may have been frustrated with my doing nothing, but that was nothing compared to when she found out the "something" I was seriously considering—television.

"There's no one who looks like you on TV!" Mom said, sputtering with fury. "You need to be a doctor or a lawyer so you can make good money and always take care of yourself!"

It was Christmastime. The tree was sparkling with ornaments and multicolored lights, and the apartment was fragrant with the smells of pine and cinnamon. We'd been in a merry mood, but Mom was so fired up about what she felt was my useless dream that she decided she wasn't going to give me my presents.

I was devastated by her reaction, but I came to understand that Mom's outburst was the protective plea of a Latina mother who'd fought her whole life to overcome poverty and low expectations. For her, law or medicine was a foolproof path to success. And I knew she had given up and delayed reaching a lot of her own goals to make sure that I could reach mine, whatever they might be.

Given how strongly Mom felt, I had things to think about. But first, I wanted to have some fun.

* * *

I'd grown up in Manhattan, but that had been a vastly different experience when I was living with my parents and required to come straight home after school. Later, in college, I would return many weekends, but there were still limits. I was a teenager with a baby face, and bouncers and bartenders weren't going to risk getting their clubs shut down because they'd let in and served booze to someone who was clearly underage.

But after I graduated from SUNY, I was twenty-one. I could go to any club that I wanted and drink if I pleased. I had a lot of time to make up for, and I was determined to make the most of it. I partied almost every night.

I would go to the World over on East Second Street on Thursdays and hit the Tunnel on Fridays. Then there were the pop-up clubs, roving parties that required a password to get in. I'd stay out until I saw the sun creep over the East River, the beats of Slick Rick and Doug E. Fresh echoing in my ears.

*Have you ever seen a show with fellas on the mic*
*With one minute rhymes that don't come out right*
*They bite, they never right, that's not polite*

My buddies had a name for our round-the-clock partying. We called it "Break Night," which meant we would dance and drink till dawn, grab a greasy breakfast to soak up the alcohol, then head to work without so much as a power nap.

One day, that epic partying got me fired. I'd gotten a job as a server at TGI Fridays, and after grabbing breakfast with friends at a diner we loved on Sixth Street and Second Avenue, I headed to work. I was more than a little hungover and probably not smelling too great either. In my wobbly state, I took the order of a customer who wanted blueberry blintzes.

We servers would rush to the kitchen, then load the plates for an entire table on these giant adobe platters that were incredibly heavy, even when your legs and hands weren't shaking from lack of sleep. Well, I spilled those blueberry-splashed hotcakes all over my customer. That was the end of that gig. I got fired pronto.

I didn't mind heading home and getting some rest, but I felt awful that I'd lost my job. I'd actually enjoyed being a waitress. I was able to boost my meager income with all the tips I got. I was constantly meeting new, interesting people, and we got to eat for free, which was a bonus since the food was pretty good. I actually liked my uniform, which, unlike the plaid get-up I'd donned every day to go to Dominican, I could decorate with pins and ribbons. Mom would frequently drop by with her best friend, Milagro, and order the day's special or a hamburger and leave me a gratuity nearly as big as the bill. But I think the two of them were secretly a bit horrified that this college graduate was waiting tables instead of doing something productive with her degree.

Not too long after I got the boot from Fridays, I saw an ad for an organization called the Training and Resources for Narcotics Council, or Trac-N for short, which trained defense attorneys. I'd taken the LSAT shortly after I'd graduated, perhaps as a backup plan or maybe just to quiet Mom's constant chatter about my potentially pursuing a legal career. I got a near perfect score, so the law was clearly something I had an aptitude for. When I was prepping for the test, the answers to the sample questions came so easily to me that I started to worry the practice exam was not true to what the LSAT would really be like. Now that I needed a job, I figured Trac-N might appreciate my LSAT performance and basic knowledge of the law. I put in an application and was hired soon after.

I was supposed to have duties similar to a paralegal, helping to compose motions and doing other administrative tasks. But in my first few weeks, I was pretty much the coffee girl. It was like being at Fridays except without Madonna's "Into the Groove" blaring in the background and change from tips jangling in my pockets. I hadn't minded refilling customers' cups at a restaurant, but I found it demeaning to be doing it in a professional office. I was also surrounded by lawyers and executive assistants who clearly took their jobs and one another very seriously, but barely looked up when I passed their desks.

Eventually my bosses decided to see what I could do beyond brewing up a good pot of coffee. I began to accompany attorneys to the infamous New York City juvenile prison on Rikers Island. The young men we met there were like kids I'd grown up with in the South Bronx, battling poverty and straining against forces trying to pull them down. Often they appeared nervous or sullen, giving one-word answers that weren't of much use to the lawyers who'd come to help them. I started to offer input on how to draw them out.

"Ask him what was going on at home that he was out at midnight when the robbery went down," I'd whisper in the lawyer's ear.

We'd learn that maybe the young man's father was raging. Or there was no home to go to because his mother had lost her job and the family had been evicted from their apartment. Sometimes we'd hear that the young man hadn't known what was about to happen but had tagged along, trying to fit in, and then when the deed was done, he'd had no choice but to run. The lawyers began to view my insights as valuable. Before long, I was doing a lot of the questioning.

My responsibilities grew. I helped to put together training programs specifically for defense attorneys who represented clients arrested for possession of narcotics. I became more immersed in the fine points of the law, like the disparate sentences for crack versus powdered cocaine, for which the former disproportionately impacted black and brown defendants. And I was able to interact with some of the most renowned attorneys in the country.

But what finally crystallized my desire to pursue law as a career was, of all things, jury duty.

I know it's an obligation many people dread, and a lot of employers, not wanting to have one less person to take care of their business, often encourage their employees to try to get out of it. But at Trac-N, jury duty was considered almost a rite of passage. My bosses, Adele Barnard, and her partner, Amy Berlin, raved about what a great opportunity it would be, giving me a window into the behind-the-scene workings of the legal system.

I headed to criminal court in lower Manhattan and got selected not for a run-of-the-mill case involving, say, a young man who jimmied a parked car for a joy ride but for a first-row seat to the trial of a man New York City's bombastic tabloids nicknamed "The Butcher of Tompkins Square Park."

His actual name was Daniel Rakowitz, and the gruesome story

that landed him in court dominated headlines for months in 1989. He'd been roommates with a beautiful woman named Monika Beerle, who was studying at the Martha Graham School of Dance. One day, when he was high on acid, Rakowitz struck her in the throat. When he realized he'd killed her, Rakowitz took a page right out of *Silence of the Lambs*.

First he cut up Beerle's body. Then, he boiled some of her body parts. Rakowitz preserved the bones, storing them in cat litter in a locker at New York City's Port Authority Bus Terminal. And those body parts he'd cooked? It was said that he put them in soup and fed them to the homeless in Tompkins Square Park.

The trial, which riveted New Yorkers, lasted for weeks. We heard testimony from dozens of witnesses. I was probably the youngest person in the jury box, and I was fascinated by the way both the defense attorneys and prosecutors pieced together their arguments like a puzzle, each question leading to an answer that fit with the next query, until they had spun a narrative that made those gathered in the courtroom believe them—or not.

Rakowitz's attorneys offered an insanity defense. But more fascinating to me was the way the prosecutors argued their case. I'll never forget the final witness. He was a homeless Latino man with curly dark hair who clearly had some mental health issues.

Sitting in the witness box, he described how cold it had been outside and how hungry he was because he hadn't eaten in a while. Suddenly a man appeared before him, offering a bowl of warm broth. He looked down. And there floating in the soup was a finger.

With that awful revelation, the prosecution rested its case. Then came the defense.

"What day is it?" the lawyer asked the witness.

"I don't know," the witness replied.

"What year is it?"

"I don't know."

"Who's the president of the United States?"

The witness didn't know Bill Clinton from John Quincy Adams. The defense attorney rested *his* case, but like overtime in a tied basketball game, the prosecutor rebounded, rising to ask one more question on redirect.

"You don't know what day it is or the month. And you've lived on the streets for many years. Why should we believe that you remember this particular day?" he asked.

"Well," the witness replied, "I don't know who the president is. But I'll never forget the day I was served soup with a finger in it."

I knew he wasn't lying.

When we finally got the chance to deliberate, some of the jurors looked at the final witness's transient lifestyle and his obviously addled mind and were dismissive of him.

"We can't pay attention to what that schizophrenic guy had to say," one argued as a few other jurors nodded their heads in agreement. I let them know I felt differently.

"He admitted that he had problems," I said. "But who would forget being served soup with a finger in it? That is one of the most honest things I've ever heard in my life."

After roughly thirteen days, when we were sequestered away from our families because the case had received so much attention, during which the twelve of us fought, seethed, but also listened, we found Rakowitz not guilty by reason of insanity. My fellow jurors were probably relieved to be done sitting in a stuffy jury room, going over sordid details, and eager to return to their normal lives. But I left that courthouse with a new sense of purpose. I wanted to play a role in the criminal justice system, and not as a juror, but as a lawyer.

There is a feeling I get when I know I'm home, when I know it's right. It's the feeling I got when I first walked into Dominican Academy. It's the comfort I felt the first time I went to a Greek step show and decided I wanted to pledge. That's how I'd felt in that courtroom, during the trial, despite the grisly crime that had brought me there.

I wanted to do what I'd seen those trial attorneys do. I wanted to be able to shape a narrative, to eke out justice for those who couldn't fight for themselves. I wanted to be where I could make a difference.

Mom got her wish. I decided to apply to law school. When I was accepted at Notre Dame Law School in the spring of 1990 as a Notre Dame Law Scholar, I left my life in New York City and relocated to South Bend, Indiana.

* * *

I studied many subjects at Notre Dame—criminal law, contracts, securities—and I had many areas I could have pursued once I took the bar exam and became a practicing attorney. But my goal was to become a prosecutor.

Like Kamala Harris, the California senator and former state attorney, and 2020 Democratic presidential nominee contender, I've been frequently asked, as a black woman, why I would be a prosecutor. Why, as a woman of color, would I choose to sit on that side of the courtroom?

The not-so-subtle subtext is that to be a prosecutor is to be a cog in a corrupt system that is effectively waging war on African American people, particularly black men.

I've been questioned about my decision at town hall meetings in Washington D.C. that I attended to get to know members of the community, and even by fellow prosecutors, when I was interviewing to become an assistant U.S. Attorney. "Are you comfortable being on this side of the aisle?" I was asked. "A prosecutor?" a neighborhood elder would say, somewhat suspiciously. "Did you ever think about being a defense attorney? Or going into civil rights?"

No. I didn't. And here's why.

I know from my own lived experience that the poor do not turn to crime because they are inherently immoral, just as most are not impoverished because they won't work. I know that they are often exhausted and defeated because they labor long and hard, yet still struggle to pay their bills and to live with dignity. And some get pulled into crime out of desperation or because they're hanging with the wrong person and are in the wrong place at a bad time.

I have witnessed in my own family the devastation that random violence leaves in its wake, the memory of my uncle Ed bleeding on a bathroom floor vivid in my mind, and I am keenly aware of how a deeply flawed criminal justice system too often traps the innocent and cripples individuals and communities of color.

But just as I am clear on the nuanced circumstances that can cause a person to veer into crime, and the racism that leads many in communities of color to be unjustly profiled and harassed, I know that to bring equity to a deeply damaged and damaging system, we have to be a counterweight on the scales of justice. We have to be members of the jury, members of the judiciary, and, yes, prosecutors too.

Defense attorneys certainly play an important role in the legal process, but the truth is they have little to no power. They are at the mercy of the court. While in the United States, you may technically be considered innocent until proven guilty, the hard truth is that those who are brown, black, and impoverished are typically viewed as less innocent than others, and they and their legal representatives are constantly playing catch-up to the evidence, arguments, and pointed lines of inquiry directed at them by the prosecution.

My father always told me it was better to work from within the system. He respected the strategizing and revolutionary tactics that freedom fighters he admired, like Martin Luther King Jr. and Malcom X and Stokely Carmichael, utilized to bring about justice and independence for their nations. But he also felt there was merit to climbing the ladder in corporate America, pulling in others behind you, and shaping policies by being part of the inside track.

I too believe in institutions, in the importance of fighting to make them work as they should as opposed to tearing them apart and then being left to navigate a world without guardrails and structure. We have a right, indeed an obligation, to bring our diverse experiences and perspectives to the top ranks of district attorney offices, courthouses, and the Department of Justice, which largely remain ivory towers.

Those beliefs shaped me as I embarked on a career in the law. And the thought of being one of the relatively few prosecutors who was black, Latino, and female didn't dissuade me. I'd grown used to being the only one.

\* \* \*

Through much of my life before Notre Dame, I was mentored primarily by women, like Ms. Lopez, who spotted my potential and pushed me forward; Sister Timothy, who never let me punt academically; and

Trac-N's founders, Adele Barnard and Amy Berlin, who said I had a sharp legal mind and became my most important references for law school.

But during my time and after I graduated from Notre Dame, many of those who gave me counsel and recommended me for opportunities were men—white men. They demonstrated that despite the very real role racism and sexism continue to play in American life, there are people who are allies, who try to do what they feel is right, rather than what is status quo, safe, or expedient.

My law professors Robert Blakey, who gave me a coveted internship, J. Eric Smithburn, who encouraged me to study law abroad, and Jay Tidmarsh mentored and prepared me for a career in the law. In 1994, with my law degree in hand, James Gillece, the president of the Notre Dame Law Alumni Association, was instrumental in getting me an interview with a man who helped me dissect and articulate the law with a precision that would eventually enable me to rack up one of the more successful prosecutorial records in the office of Washington D.C.'s U.S. Attorney. Jim remained my mentor and champion until he died.

Robert Bell was an African American judge who'd earned his law degree at Harvard Law School. A veteran of the civil rights era, he was the only judge to have sat on every court in Maryland. When I went to see him about possibly becoming his clerk, Bell was on the state's appellate court, presiding over civil and criminal appeals in Baltimore.

Our first encounter didn't initially bode well. During my interview, as I recall, he asked me my position on a case where a thirteen-year-old girl had had sex with a nineteen-year-old man "consensually." I immediately replied, "Well, she can't consent. She's a child at thirteen years old. He's a man." Judge Bell calmly replied, "But she looked older, he thought she was older, and she wanted to have sex. It wasn't the first time she had sex. They were dating. This isn't some stranger that pulled her into a dark alley off the street. Why is that rape?" I stammered, "Because it's statutory rape in Maryland and quite frankly all over the country. There are reasons why the law is in place." Judge Bell replied, "Maybe we should change the law. We can do that as the highest court in Maryland." I couldn't believe what he was saying. I was sitting in the chambers of this esteemed jurist whose career I'd carefully studied in preparation for our meeting, and we ended up in a screaming match. A few days after, I received a call from Judge Bell. He said, "You passed."

"Pardon me?" I said, perplexed.

It had been a test. Judge Bell had deliberately raised a provocative topic to see how I would react because he didn't want a yes-person. Rather, he wanted somebody who could craft an argument and then vigorously defend it. Once his words sunk in and my heart slowed down, I was pleased, then elated. I was going to be a clerk for an appellate judge.

In the year that I worked for Judge Bell, I researched legal precedents to prepare briefs that became the frameworks for his opinions. I was taking the theoretical, textbook learning that I had tackled in law school and was applying it to real, black-letter law that would impact actual lives. I honed my writing skills. And there were many nights debating esoteric points of law. But even as I was advised to pursue a variety of experiences before I decided where to settle, in the back of my mind, becoming a prosecutor was still my ultimate goal.

Once my clerkship was over, I briefly shifted to a medical malpractice firm. It handled largely civil cases, with the few criminal matters that it dealt with mostly involving clients who had the wealth to hire former prosecutors to get them off the hook for securities fraud and other white-collar crimes.

Again, I did my due diligence, trying out different aspects of the law as my mentors and professors advised, both to beef up my résumé and to help me make a well-informed decision about what I wanted to pursue. But in that firm, in that space, my inner compass told me I was moving in the wrong direction. It didn't feel like home. I'd been educated by nuns who'd devoted their lives to education and religious service, and I'd grown up with a mother committed to activism, so it just didn't feel right to be immersed in a corporate environment where legal services were afforded to the highest bidder. I was feeling restless and dissatisfied.

I applied to the Department of Justice Honors Program hoping to finally fulfill my dream of working full-time on criminal cases. But when the lawyers there saw my résumé, they instead offered me a job in the Antitrust Division. It wasn't what I wanted, but it was a way in. And the Antitrust Division was considered the department's crown jewel.

The mission of our division was to try to prevent monopolies, and whatever case you worked on, you had to become an instant expert. I'd done well in courses on securities in law school and I was getting to put that knowledge to good use, as well as gaining a thorough understanding of a whole range of industries. I met incredible attorneys and friends there whom I still admire, like Dando Cellini, an attorney, who like me, loved to stay at wonderful hotels and try new restaurants while on assignment, willing to pay a little extra over the government-allowed per diem, and Ali Ramadahn, an avid hunter who hunted not for sport but only for food and made the best venison in D.C.

But I wasn't doing the kind of advocacy that I craved or working on cases that got me into a courtroom.

When I sought out more stimulating assignments, my bosses fed me the old adage of my being a victim of my own success, which I've always felt is a not-so-slick way for people to keep you on tedious assignments that others don't want. But one new task that I enjoyed was recruiting at law school fairs, particularly for hires that would increase diversity in the department, which had very few attorneys of color.

At one of those events, I found myself sitting next to Wilma Lewis. She was a dynamic black attorney who also had Puerto Rican roots, having been born in the town of Santurce. She's now the chief judge of the district court of the U.S. Virgin Islands. But when I met her, she was the first woman to be the U.S. Attorney for the District of Columbia. We struck up a conversation.

"I'm in the Antitrust Division," I said, careful to be diplomatic so that I didn't appear to be badmouthing my boss to someone who could potentially become a new one. "I've done well there, but I always thought I'd be in the Bronx or Manhattan District Attorney's office. You know, having an impact on a more local level."

She gave me her business card. "You should call my office," Lewis said. "I might have to steal you away."

A few weeks later, I was scheduled to go in for a series of interviews. I was being considered for a position in one of two units, one that focused on misdemeanors and another that handled appeals.

I really wanted to be in misdemeanors, which would mean being in the courtroom virtually every single day, dealing with cases of prostitution and low-level drug busts. But I'd already tried a federal criminal case when I'd worked at Miles and Stockbridge with Dick Bennett, who later became a federal judge. I'd clerked for Judge Bell and handled antitrust matters. That experience gave my résumé a luster that was considered a bit too highbrow for prosecuting petty crimes.

I guess the thinking was, "How is this chick, who went to Notre Dame, was well regarded in the Antitrust Division, and who Dick Bennett wrote a letter of recommendation for, going to feel handling cases involving street-corner exchanges of crack in southeast D.C.?"

Looking back on it, I didn't have much experience in that particular legal area. And I didn't approve of the unit's "broken window policing," the practice of coming down hard on the most minor of infractions to supposedly prevent an escalation to more serious crimes. I thought it was draconian. If I'd gone into misdemeanors, I likely would have raised more than a little hell.

Still, I totally didn't want that appellate gig. In that division, you had your own office or one you shared with another colleague. The misdemeanor unit was a bullpen, where you had a desk in the middle of the clamor. That's where I wanted to be, in the heart of the action, not walled off in a claustrophobic office. In the end, however, I didn't have a choice. I had to interview with both and see which one wanted me more.

The day I was to interview with the appellate division, I showed up ready for back-to-back chats with several lawyers. I wasn't earning a lot of money at the time, but I'd gone to Ann Taylor and bought a sleek black-and-white houndstooth pantsuit that I thought made me look professional but stylish. I felt good as the day progressed. Before my final interview with the Appellate Division's chief, John Fisher, I grabbed a seat in the bullpen, where I waited.

Wilma Lewis had a staff full of black women. One of them, Brenda Baldwin-White, was in the front office with another lawyer, Deborah Long-Doyle, a light-skinned black woman with short, curly red hair, freckles, and red glasses. After I'd sat down, I noticed Brenda looking at me. She walked over, introduced herself, and asked me to come into her office—one of the few set aside for her and other top staffers—for a minute.

She shut the door and perched on the edge of her desk. "You're about to meet with John Fisher," she said. "That means the job in appellate is almost yours. But you won't have a shot if you walk into Fisher's office wearing pants."

"Excuse me?" I asked, not sure what I was hearing.

"He's very old-school, the most old-fashioned guy in this building," she continued matter-of-factly. "He won't hire a woman in a pantsuit. You're going to have to change."

"Change?" I asked, stunned. "I don't have an extra outfit in my purse! How can I change? And not to be disrespectful, but I really don't want to be in the appellate division. I want to go to misdemeanors."

Now, speaking to me like she was dealing with a petulant four-year-old, Brenda calmly explained that a placement in the appellate division was a prize that not a lot of black women got, and it would be a great addition to my résumé. She looked me up and down.

I was five-six and weighed about 110 pounds. Brenda was roughly the same size. She stood up.

"Here, I'll give you my skirt," she said. "And I'll put on your pants."

I felt like I was having an out-of-body experience. Take off my pants? Wear a stranger's skirt? For what? To get a job with a chauvinist in a division where I didn't even want to work?

"I'm not doing it," I said, literally digging my three-inch pumps into the worn government carpet.

Brenda picked up her phone to call in reinforcements. She explained the situation, never averting her gaze as she spoke into the receiver. A couple of minutes later, Deborah and June Jeffries, a homicide prosecutor who later became the head of the division, walked in.

"Take the skirt," June said. "Trust us. You want this job. Don't let a pair of pants keep you from it."

Brenda, June, and Deborah would one day co-host my baby shower when I was expecting my son, Gabriel. Years after that, they'd travel from Washington D.C. to Westchester County in New York to surprise me at my fiftieth birthday party. But that day was the first time I'd ever laid eyes on these women. This situation was crazy.

But I also knew that they were three powerful sisters working in the U.S. Attorney's office. For them to propose something so absurd, it had to be important that I get that job, for them as well as for me.

I shimmied out of my pants and shimmied into Brenda's black skirt. Then, with my borrowed outfit, and more than a little bit of attitude, I met with John Fisher. I got the gig.

Should I have had to put on a skirt to get hired? Of course not. Did I think it was ridiculous? Absolutely.

But if you want things to change, you have to be on the inside to change them. That may mean knowing when to hold your tongue or temporarily dealing with some absurdity. I gained so much from working with John Fisher, who is now a judge on the D.C. Court of Appeals. He was smart as a whip and definitely didn't underestimate my skills or intellect because I was a woman.

It was also just a fact that Washington D.C. was Southern in its sensibility, and even in 1997, there was a certain old-fashioned decorum that people abided by. The no-pants-allowed rule wasn't just an outdated requirement imposed by Fisher. Women couldn't wear pantsuits in front of juries either, at least not if they wanted to win. You had to wear a skirt and blazer, or the judge and jury would be too busy tsk-tsking your trousers to listen to your argument. It was silly, and thankfully now it's changed, but it was a concession you had to make if you wanted to succeed. And I did.

The U.S. Attorney's office in Washington D.C. handled a mixture of local and federal crimes. We argued cases before the D.C. Court of Appeals, and the US Court of Appeals for the D.C. Circuit, the number two court in the country, where Brett Kavanaugh, the controversial and newest U.S. Supreme Court Justice, once sat.

When you are arguing civil and criminal appeals cases before men and women who have a clear shot at sitting on the highest court in the land, you have to be damn good. You face a panel of three judges or more, by yourself or as part of a team. The cases are complex, and you have to become an expert very quickly, ready to answer any and every question that might be tossed your way.

An African American attorney named L. Jackson Thomas helped show me the ropes. He was mercurial, but he'd been in the appellate division for years, and he paid particular attention to lawyers of color, giving us guidance during our time there. His tutoring was much appreciated because the job was incredibly stressful. Before every appellate court argument I gave, I vomited and had diarrhea.

No matter how much encouragement I got from Jack and others, who reminded me that I knew more than anyone else about my case, I always feared that the judges and others in the courtroom were smarter than me. I found the experience to be incredibly intimidating, and I never stopped feeling frantic inside.

You'd write and rewrite your briefs. And while today you submit them electronically, back then, you had to go to court and drop them off in person. A deadline was a deadline, mandated by the court rules. If you missed it, the defense could get the case dismissed. Many a day, you'd be sprinting down the street, trying to get to the clerk. Or, when you had a hearing, you'd pace the hall, waiting for the courtroom's doors to open.

There was a limited window of time in which to present your case, and there'd be red, yellow, and green lights warning you how many minutes you had left and potentially distracting you from your next point. Sometimes a representative from the U.S. Attorney's office would be there to watch how you performed. You never knew what questions you'd be asked, and the judges would fire them at you, fast and furiously. You had to be articulate, cool, and clearheaded in front of these black-robed jurists who looked like they'd stepped down from Mount Olympus. I found my time in the appellate division to be one of the most unnerving experiences of my life.

But it also showed me what I was capable of. And the debate skills I honed there, along with the ability to perform under pressure, would help me when I later prosecuted rapists and drug dealers, as well as when I got into heated discussions in front of millions of viewers as a legal analyst and anchor on TV.

After five or six months, I rotated out of the appellate division. I took on some misdemeanor cases before moving on to felony offenses involving guns and narcotics. I'm sure some of my colleagues looked at that shift as a bit of a demotion.

But I'd become a damn good trial lawyer. And at that moment, in my new assignment, I was finally able to take all the experiences I'd gathered in my life, during my childhood on the streets of the South Bronx, in my time absorbing information in the Antitrust Division and during my arguments in front of appellate judges, and use them in the way I'd wanted—to prosecute criminal cases committed by and against working-class people. My people.

To do that effectively, I was going to have to be on the streets, in the churches, and in the courtrooms where lives were lived, crimes were committed, and justice was meted out. I had to be where the action happened. I was ready.

# CHAPTER SIX

# THE ROOM WHERE IT HAPPENS

In neighborhoods like the South Bronx, Harlem, and southeast Washington D.C., community activists play a vital role. They can agitate for change, patrol a dangerous block, and create an outlet to channel and convey community concerns and rage.

But if you're going to change a law, you've got to be a lawmaker. And if you're going to decide what a young person gets charged with, if he or she is charged at all, you need to be a prosecutor. To be on the minds of jurors as they deliberate, to tip the scales when a judge decides what sentence to hand down, you have to be on the government's side of the table. You have to be in the room.

Like Eric Holder, the U.S. Attorney General under President Barack Obama, Wilma Lewis made it her mission to increase diversity in her office. Just as she'd given me her card at a job fair, then asked me to give her a call, Lewis actively recruited many other African American lawyers. She wanted to make sure the men and women pursuing cases that could upend or salvage lives mirrored the population that they served, and could relate, even if only a little bit, to their reality. I used to joke that her front office looked like the set of *The Color Purple*, the film with an all-black cast that was based on the Pulitzer Prize–winning Alice Walker novel. It was the most diverse space I'd ever been in professionally since becoming a lawyer.

I would spend five years there, enough time to take full advantage of one of the best features of the job: the chance to rotate through its various sections. After my stint in the appellate division, I did round-robin duty presenting to the grand jury, pursuing cases involving guns and drugs, prosecuting perpetrators of domestic violence, and investigating sex crimes and homicides.

Those experiences influenced actions and stances I would take in all aspects of my life. I sit on the board of Safe Horizon, a national organization dedicated to assisting victims of violence, because of the men and women and children I met and fought for as a federal prosecutor. And as an assistant U.S. Attorney, I gained a far deeper understanding of how communities of color are treated and viewed by law enforcement, which I carried with me when I became a legal analyst, network correspondent, and talk show host, jousting with others about policing and the workings of our legal system.

During my time in that office, I also believe that I played a role in shaping the views of some of my peers. Not many of the attorneys I worked with had a background quite like mine. My working-class childhood spent in tenement walk-ups and housing projects, my own brushes with violence when I saw my uncle stabbed and the aftermath of my best friend's father being murdered and having members of my own family behind bars, gave me a unique perspective that I shared with the professionals around me. I felt like the people in the community, both those who were victims and those who preyed upon them, were a part of me, and I was a part of them.

There were more official efforts under way to forge the kind of connection I naturally felt to the folks we were supposed to look out for.
Community policing came on the scene in the 1970s and hearkened
back to the old-fashioned idea of the cop on the beat, when officers
knew the names of the people in the neighborhood they patrolled. It
put an emphasis on partnering with citizens to stop crime before it
happened, rather than having an "us versus them" relationship and
bringing down a hammer after the fact.

The U.S. Attorney's office in Washington D.C. adopted a similar
approach. The city was carved into quadrants, and a prosecutor was
assigned to each, making him or her responsible for the cases that
occurred in the area. The thought was that this could be a more ef-
fective way of prosecuting because people in the neighborhood got to
know not only the local cops but the attorneys who would be pursuing
the perpetrators of crime. And your very presence might deter someone from taking part in something they had no business doing.

The prosecutors got a 360-degree view of the community and the
law. Monday might be spent in the office, sitting with police officers
and reviewing the cases that came flooding in. Then Tuesday, you'd
drive through the neighborhood with detectives on a ride-along.
Wednesday, you could be in court all day, then rush to a neighborhood church or school to attend an evening town hall meeting. And
the next day you might head out on your own to knock on doors, like
a politician, to meet your constituents.

I was assigned the sixth district, a particularly violent pocket of
D.C. Going to witnesses' homes was more than a meet and greet—it
was a necessity because there were many challenges that prevented
them from coming to me.

Our office was at 555 Fourth Street Northwest, nicknamed "Triple Nickel" by the lawyers, witnesses, and suspects who streamed

through. It wasn't located in the heart of the neighborhood, but there were always folks keeping watch on who came and went, and if someone spotted a potential witness leaving our office, that could get them labeled a snitch and put them and their families at risk for retribution.

Even for those who weren't worried about being seen, just getting to our office could be arduous. Many of the working-class folks who called D.C. home didn't have childcare. Few could afford to take off work and lose a day's wages. Public transportation wasn't very reliable in their part of the district, and trekking back home after dark could be dangerous. Even if they had a car, filling the tank cost money that might be better spent on groceries or to pay the water bill.

So the best way to connect was to go to their homes. Unlike several of my colleagues, who would venture out only when they were accompanied by one of the officers, I preferred to go alone because a cop showing up at your front door was like waving a flag to every ne'er-do-well on the block. You could forget anybody talking to you after some incident had gone down.

I can't emphasize enough how much those personal visits meant. Being out in the community meant that I was able to build relationships with residents who, often with good reason, did not trust a system they believed persecuted innocent black people almost as much as the drug dealer hawking drugs a few feet from their front doors. I believe conviction rates around the country would be higher for crimes in which people of color are attacked if they felt more comfortable coming forward.

During these visits, I would walk up on a stoop and recite some version of a script that went something like this:

"Hello, ma'am, I'm your community prosecutor. I understand there was a shooting the other night. I was sorry to hear that a bullet shattered your living room window while your family was sitting inside. My office has been calling you, leaving messages, but we haven't heard back."

I would tell people who often felt they had few choices in life that in this instance, they had one. "Andre and his crew have this whole block on lock," I'd say, "and if you don't testify against him, he's going to continue to shoot up your neighborhood. He's going to keep standing in front of your building telling you when you can and can't come out of your own home, or trying to get your son to sell for him."

Finally, I would let them know that I was on their side. "I'll be with you every step of the way. I can protect you. I can even relocate you and your family if you feel threatened."

Of course, there was little reason for anyone to believe what I was saying when they saw me on their block for the first time. That's why the community-engagement philosophy mattered so much. We had to keep coming back, to be seen chitchatting on the corner, at the barbershop, and in the hair salon. We had to become a part of the community.

Some of my fellow prosecutors weren't as comfortable being out and about. They'd go on a ride-along with detectives, who made them feel safe because they were carrying guns, but they weren't going to just show up at a potential witness's home by themselves. They preferred to sit in the office and work the phones. I on the other hand was eager to hit the streets. I wasn't afraid to walk through a frayed neighborhood that others wrote off as dangerous. I wasn't nervous about knocking on the doors of strangers and then going inside to sit on couches sheathed in sticky plastic to protect their store-bought sheen. The South Bronx was a lot like D.C., just 238 miles away. I'd been walking through those streets, entering those houses, and sitting on those couches my entire life.

I started going to the local elementary school, Amidon-Bowen Elementary, once a week to read to the students. And I became like a big sister to some of the neighborhood kids. One boy I met, Naquaran, had a very abusive father, and there were times I let him crash in a spare room at my home. All these years later, we are still in touch. He is a kind, smart soul who has had a difficult life and made some bad decisions. I hope eventually, with my help and the guidance of others, he will start living up to his potential and stay out of jail for good.

The cops I worked with worried about me. "Did I hear you were down in the hood by yourself the other day?" one would ask me incredulously. "You can't drive your green BMW down there, AUSA [Assistant U.S. Attorney] Hostin!"

I explained to them that the street reminded me of Fordham Road in the Bronx, or 125th Street in Harlem. I would tell them that the apartment I'd visited the day before looked a lot like the place where my cousins Sean and Tyvee used to play. "I'm good," I'd say.

Once again, I was straddling two worlds, and as always, having one foot in each gave me insight into both. I understood the unease many people in the neighborhood felt about cops and courtrooms. And I made the troubling observation that some in law enforcement, including prosecutors, were afraid of the very people that they were supposed to protect and serve, the children and elders whose side they were supposed to be on. I didn't understand how helpful you could be if you were warily circling the folks whose cooperation you'd need to solve a case or to keep the peace.

When it became clear that they couldn't dissuade my solo trips, the officers I regularly worked with stopped bringing it up. And eventually my comfort in the community led them to rely on me to be an interpreter of sorts, though sometimes I was as in the dark as they were.

One morning, a group of prosecutors and agents were huddled at Triple Nickel, listening to a tape of an undercover sting operation. Now, true, I'd grown up immersed in the rhythms of the Bronx. But I had moved away at the age of eight, and after attending private school, going to college, and attaining a law degree, my life had in some ways gotten decidedly bougie. I vacationed on Martha's Vineyard and attended cocktail parties with federal judges. I was more familiar with the Latin I studied at Dominican for four years than the slang spoken on the street.

But I guess I was all they had. They asked me to come listen to the tape—then asked me to translate. I had no idea what the guys caught in the sting were saying. But while it was funny they thought my Bronx roots could help them decipher the street code, it was also a little unsettling that they were so out of touch, they couldn't figure out what they were listening to. All I had to do was call a cousin (which I did) and I would have the translation. They had no connection.

More often, I was in sync with the people I came to know in my district. I felt much more at ease and full of purpose in those crowded town halls and tiny homes than sitting in an office studying documents, meeting with turbine manufacturers, or listening to speeches at a fancy luncheon. Though I could do that too. I'd had a life in which I'd perfected being a chameleon, adjusting and blending in, but some spaces felt far more like home than others. And I know that one of the reasons I never lost a case was because the people I advocated for saw how comfortable I was and grew to trust me.

I remember dropping by the home of a neighborhood elder to wish her a happy Thanksgiving. She was cooking up a soul-food feast.

There were pots of chitterlings and collard greens on the stove, and she'd prepared banana pudding topped with vanilla wafers. She and her family asked me if I was hungry and graciously offered me a plate.

I was happy to sit down and eat, but in all honesty, chitterlings aren't my thing. Knowing it would have been insulting not to partake, I doused them with hefty splashes of hot sauce—not the Red Hot kind, but the real deal from Louisiana—and dug in. I also asked for some of that banana pudding.

I knew that moments like that made the people I wanted to help feel as though I'd shown up for them and made them want to show up for me when I needed their perspective, their insights, or their statements to put together a case.

* * *

I wasn't on the job long when I began to notice something deeply disturbing. More times than I could count, police officers I worked with handed me written confessions accompanied by photographs of African American men's battered faces.

In the booking photo, or a picture snapped when the suspect was in a lineup, I would see a busted lip, a black eye, or a ripped shirt. I always noticed. I always questioned it.

"What is this?" I would ask.

I would listen as they sputtered or outright lied; then I'd calmly let them know, "Well, we can't use it."

I wasn't the only attorney to see such photos. But unfortunately, too many of them, instead of questioning the veracity of an accompanying confession, would try to get the suspect to plea. If he or she was a repeat offender, poor, or both, which was often the case, many times they'd take the deal and wind up enmeshed in the system. That's not to say those who said they were guilty in exchange for a lighter sentence were actually innocent, because a lot of them weren't. But opting for a legal shortcut, without asking enough questions, didn't feel right to me. Especially a confession that had been coerced. A disheveled suspect in a photo, an account that seemed a little too pat and perfect, set off my antennae. And I was proven right.

There are ways to make sure that justice is done without rewarding bad police tactics. I put the officers I worked with on notice about behaviors that I was not going to co-sign or tolerate. I'd like to think that they thought twice about roughing up someone in custody after dealing with me, if not because it was simply wrong to do, then because they weren't going to risk one of their cases being tossed and never having its day in court. If my hard line kept them from putting their hands on someone, that was a step in the right direction, even if they ultimately held back for selfish reasons.

That's the importance of being part of the system. Being inside means that you are intimately familiar with all the ways that it is broken, and it gives you an edge in figuring out how it can be fixed.

This matters to me because one of those battered men could have been my cousin Sean, Donnell, Ronnell, Jeff, Travis, or my uncle Ed. Like so many families, there was barely one degree of separation between those like me walking the straight and narrow, and others who collided with the law.

That's the tension in many families, brown, white, or black, wealthy, poor, or working-class. But those intrafamily divisions can be more fraught among people of color, for whom opportunities are more limited and scrutiny is more severe. Those who make it are often pressured to reach behind and lift even those who do not want to climb. And if they are not careful, the ones who've done well may see their success snatched back and end up tumbling off the ladder.

* * *

Paying attention and listening to every single detail makes the difference between a good lawyer and a great lawyer. Intuiting is particularly critical when working with victims of sex crimes.

When I worked in the sex crimes unit, a case came up involving a high school teenager who was allegedly in a sexual relationship with her high school's football coach. She was an African American girl, only fourteen, and he was forty. She didn't want to cooperate with the investigation, but some of my colleagues, knowing the success I'd had connecting with people in the community, thought that she might be more willing to talk if I got involved. So the investigation was handed to me.

When I first met her, I was struck by her beauty and poise. She sat in my office, in a chair between her father, who worked for the fire department, and her mother, an executive assistant for a city official.

While her mother cast her eyes downward, her hands fidgeting in her lap, the girl's father was visibly irate. I understood why he was upset. Still, his rage seemed over-the-top in a way I found unsettling.

The victim said that she still didn't want to cooperate. "It wasn't rape," she said in a near whisper.

I had dealt with many disturbing cases in that unit. I tried to get her to meet my gaze.

"You may feel like you're in a relationship with him," I said softly, "but you are too young to consent under the law. So it's nonconsensual."

We proceeded to press charges. The coach was arrested, and I pre-pared for trial. But the case continued to disturb me, and not just because a teenage girl had been taken advantage of by a forty-year-old man, which was horrific enough. In the numerous meetings I had with the girl and her family, preparing the case, her father repeatedly said something incredibly bizarre.

"He needs to garden in his own backyard. He needs to garden in his own backyard!" What father, talking about the sexual abuse of his child, would speak of her abuser in that way?

The girl also had a godfather who she was very close to and who would sometimes accompany her to my office in the weeks leading up to the trial. I'd finally convinced the young lady to testify, though she remained reluctant. Her godfather thanked me for what I was doing for her and praised how caring and thoughtful he felt I was being. Then, he said, there was something else he wanted me to do.

"You need to look deeper into this, because a lot of stuff is going on that people aren't talking about."

"Like what?" I asked.

"Just look deeper," he said. "There's a lot of secrets in this family. A lot of secrets."

When we finally had our day in court, the coach was convicted and sentenced to prison. Afterward, as I did with every case, I asked the family to come see me just so I could check in on how they were feeling and to let them know that even though the trial was over, they could always come to me to talk about any concerns. When they arrived, the tension in the office was so thick, it made the air hard to breathe.

The mother, always unsettlingly quiet, now seemed numb to the point that she was barely present. The dislike the godfather had for the girl's father was etched into every crevice of his broad face. The dad, however, was almost giddy.

He punched his fist in the air, as if he'd filled out a perfect March Madness bracket rather than just seen a man sent to prison for having sex with his underage daughter. He gave me a hug, sat down next to his daughter, then placed his hand on her thigh, close enough to nearly touch her crotch.

"Thank you for all your kind words," I said slowly, looking at him directly. "If it's okay with you, could I have a few words alone with your daughter? It's something I'd like to do to make sure she's okay with the process."

The other adults in the room filed out. I waited a few seconds after they closed the door. Then, I didn't feel there was any point in circling around the obvious. She knew me. I knew her. And there wasn't much time before her parents would be knocking, asking to come back in. I asked her, "How long has your father been abusing you?"

She burst into tears. "For as long as I can remember."

I filed charges the next month, this time against the young lady's father. In our initial conversation that day in my office, she'd told me she was willing to tell the truth on the witness stand about what her father had done. But unfortunately, when we finally went to trial, she and her mother, who knew about the abuse, both flipped the script, saying he'd done nothing.

I was able to salvage the case because the family had gone to counseling, and some of the terrible details of what the girl had endured were revealed in the presence of her godfather, who took part in some of the sessions. His being there negated the standard confidentiality between a therapist and patient, allowing me to subpoena the notes and reference them in court.

U.S. Attorney's Office swearing in with then U.S. Attorney, Wilma Lewis

Presented with Special Achievement Award Prosecuting
Sex Crimes by then U.S. Attorney, Wilma Lewis

Paloma and Gabriel

Pregnant
with Gabriel

High school reunion, Dominican Academy

Sunny and Manny's wedding, August 1998

Sunny with her mother, Rosa Beza, in the Bronx

Sunny with her grandmothers for her First Communion

Nannie Virginia's basement apartment, the first time Manny met her

Sunny in the hospital with Gabriel,
the day of Paloma's birth, 2006

Paloma at two years old

Gabriel at two years old

Sunny as a baby

Sunny, Jeff, Sean, and Nannie Mary

Uncle Ed, Tyvee, and Sue Bell

Jeff and Sunny

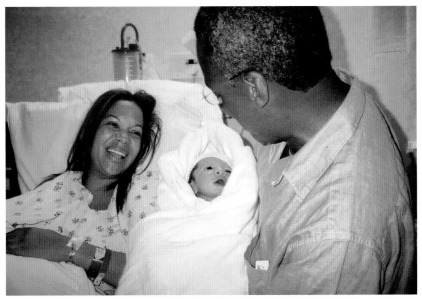

Sunny's father in the hospital holding baby Gabriel, 2002

Nannie Virginia

Gabriel's baby shower

Gabriel's baby shower *(left to right)*: Karla Dee Clark, June Jefferies, and Deborah Long-Doyle. June and Deborah are friends from the U.S. Attorney's Office who donated maternity leave and convinced Sunny to change her pants for her interview.

Sunny and Manny's
wedding, August 1998

Sunny and Manny on their first date

The Romero women

In front of the D.C. U.S. Attorney's Office

Sunny with her mother

Gabriel's first smile

Casual day at the Court of Appeals

Sunny with Marshall, another law clerk

Best friends: Sunny with Regina Jansen, Tamika Tremaglio,
and Stephanie Rawlings-Blake

Sunny with Supreme Court Justice Sonia Sotomayor at the Court of Appeals

Sunny on the beach with Manny, Gabriel, and Paloma in Turks and Caicos

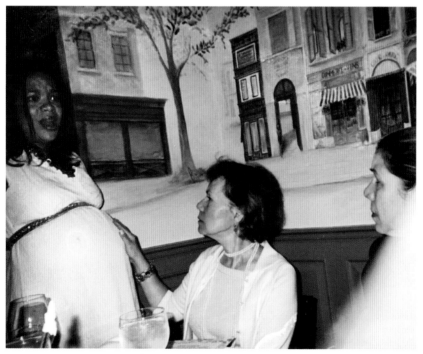

Pregnant with Paloma, 2006

Mom with sisters, Carmen and Inez

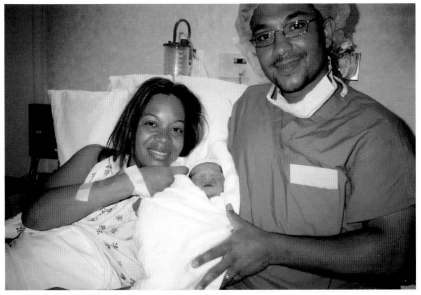

Sunny and Manny in the hospital after Gabriel was born, August 15, 2002

Paloma playing outside

Sunny with Paloma

Sunny as guest co-host on *The View* on Friday, May 9, 2014.
*(Left to right)*: Sunny, Barbara Walters, guest Bette Midler,
Sherri Shepherd, and Whoopi Goldberg

Sunny and Barbara Walters

Sunny and Colin Kaepernick

Sunny covering the Cheshire home
invasion case in Connecticut for CNN

Sunny with *The View* executive producer Bill Geddie

A jury ultimately found him guilty, but he received a relatively short sentence. Sometimes you get justice, but not as much as you would like. And sometimes you have questions about whether you got justice at all.

Once, when I was in the domestic violence unit, a woman came in with photos showing her bruised face. The man who she claimed abused her was prominent on the Washington D.C. social scene. She had not sought medical attention so there were no medical records to support her claims. When he came in for questioning, he was accompanied by his lawyer, and it became a push and pull of he said, she said. He painted his accuser as a gold digger who'd stolen from him, filed phony suits to extract more money, and had behaved so erratically that he'd been forced to take out a restraining order against her. The police believed him.

Domestic violence is a scourge in our society and women are often abused by men who they believe love them. But pictures of a battered face are not enough proof of assault. He argued that those injuries were self-inflicted and that she often attacked him and he had to act in self-defense. As the prosecutor, I had to make the call about whether to proceed with the case or to refuse to take it. This was one of the few times in my career when I couldn't decide.

I told the division chief that I was confounded. She looked at the evidence and said what I was thinking but didn't want to necessarily admit: "We can't prove it."

When I called the former couple into my office separately to tell them we wouldn't be filing charges, he gave his lawyer a high five. She left my office sobbing. I felt awful, that I had let her down. I believed he had done wrong but that there wasn't enough evidence there to prove it. That was a bad day. But not my worst.

There was another day, another case, when the fates of a group of black children was in my hands.

A teenage girl said she had been hanging out with a group of boys in an abandoned apartment when they gave her something to drink and had sex with her without her consent. They were all about sixteen or seventeen years old and we decided to do what was called a "U Ref," meaning we lowered the charge from a felony to a misdemeanor.

Still, in Washington D.C., even a misdemeanor sex crime meant you must register as a sex offender, which is death to your future. Again, these were black boys, who already had so many strikes against them by virtue of the color of their skin.

The accused young men wanted to meet with me because I was well-known in the community. When their parents called to ask if they could come to my office, I told them they needed to contact the public defender, get legal representation, then reach out to me. That was the appropriate way to go forward.

When their lawyer did finally call, the boys' story of what occurred was of course quite different. Yes, they had gotten together to hang out and drink, but they claimed the girl had willingly had sex with each one of them. Now, after the fact, she was saying it was rape. I acknowledged that the details of what happened were murky, which was why we were charging them with a misdemeanor instead of a more serious offense.

"If they plead, at most, they'll be in jail for a couple of months," I said.

"Yes," their attorney agreed. "But they will have to register as sex offenders."

My hands were tied. There was nothing I could do about the D.C. law and its requirement. And it was their word against hers.

Insisting it wasn't rape, the young men wouldn't enter a plea, and I took the case to trial. She testified against the boys. The jury didn't deliberate long. The verdict? Guilty.

When the teenage boys were led away, one of them, a tall, handsome young man, turned to look at me. He shook his head as if to say I had gotten it wrong.

That was the worst day. It's been twenty years, and the look on his face is etched in my memory so clearly it is as though I am seeing him right now, standing in front of me. I wonder if the jury made the right decision, if there is anything I could have changed. I question whether justice was done.

Given the conflicting stories the boys and the girl told, and the uncertainty I felt in my core, I can say that I would feel the same doubts, the same pangs of guilt, if the accused young men had been purple. But the fact that they were young black men makes the feelings somehow sharper, because by virtue of their race, I know that they already had everything else against them. That's one of the most difficult parts of being a prosecutor, because we know that people of color are arrested and prosecuted and incarcerated at rates that far exceed those of their white counterparts.

When we discuss the law, we talk about winning and losing as though those are equal outcomes, but what is gained can be immeasurably more than the alternative: liberating a city block from a young man who terrorizes its residents or freeing a young girl from the predatory behavior of an abusive father. Or, the loss can be incalculable, reverberating through a life long after a sentence is served, occasionally robbing the wrong person of their freedom and their potential.

When people question my decision to become a prosecutor, I can tell them that on my watch, when I did pursue cases against people of color, it was because I felt they had merit and the evidence to back them up. They were never based on hazy descriptions, questionable confessions, or possibly biased assumptions.

Despite my doubts about the cases involving the young men, and the woman who accused her ex-boyfriend of abuse, I felt far more certain about the many others that I prosecuted. I know that I made equitable decisions, and I truly believe that if there were more prosecutors of color, the system would be more just because we bring our life experiences and our perspectives to bear.

* * *

All those crimes, all those questions, can take you to a dark place. You definitely see the worst of humanity. But you also see the best.

There were grandmothers who came forward about a violent drug dealer, risking their own safety to protect their neighbors. And there were activists who, after working all day, would mentor young people and give me guidance as I learned the rhythms of their community. There are so many people, often unsung, often at great risk, who try to do the right thing.

Despite the bleak crimes I had to confront, I felt worse when I came home from a corporate law firm where I dealt with jerks making boatloads of money. Then I would think, "What am I doing with my talent? Is this what my parents and grandmothers and Ms. Lopez and Sister Timothy envisioned for me? Helping the superrich take more money from someone else?" That didn't make sense.

Seeking justice for the most vulnerable in our society—children, the elderly, those who are struggling every day to etch a better life—is God's work. There's nothing more important than trying to protect them. And whatever the outcome, just being in that fight means you're on the right side every time.

I would not have been able to advocate for the young girl who had been abused if I hadn't been in the room. I wouldn't have been able to gain her trust if I hadn't remained in touch with the community she came from. All the stereotypes and prejudices that women, and people of color, have to face can be devastating obstacles to navigate. But it was those very aspects of myself, and my ability to see my reflection in the eyes of the people I served, that has helped me to successfully fight on their behalf.

I am not naïve. When I engaged in voir dire, quizzing the randomly selected men and women who could potentially deliberate a fellow citizen's fate, I was under no illusions that they would magically shed their biases before taking their seats in the jury box. And I'd scrapped with enough judges to understand that they were not automatically objective simply because they'd taken an oath and donned a black robe. But with the lives of so many black and brown men on the line, I felt it was important to have someone like me sitting on the side of the prosecution, someone who saw not just a suspect but a soul; not just a delinquent but somebody's son.

There may have to be a lot of patch jobs on the way to permanent solutions. I knew then, and know now, that no single person can on their own transform a system that has been methodically built and manipulated for generations. However, I could take it upon myself to apply some of the elbow grease needed to make the wheels of justice turn a little more smoothly. I just had to be in the room.

# CHAPTER SEVEN

# MOTHERHOOD

When I was sixteen years old—roughly the age my mother and father were when they met, got pregnant, and married—my parents got a divorce.

People often say when their mothers and fathers break up that they had a feeling, saw warning signs, heard something in their mother's voice or in the silence that loomed whenever the family was together. But there was none of that for me.

My parents had arguments. Bad ones. I remember my father going out with his friend Butchie on a Friday night and not coming back until Sunday morning. They claimed they were at a Muslim retreat. My father even had on some sort of headdress, and he was wearing only one shoe. My mother hit him over the head with an African statue we had in the house, leaving a huge gash in his head. But a few days later I heard them laughing again. Meanwhile, my mother had serious bouts of depression and drinking, even a stint in rehab. But weeks after she got back, I heard them laughing again. This time, though, the laughter never returned to our apartment.

I was away at SUNY Binghamton when I found out. I don't remember there being any big discussion. My parents didn't drive up
to campus to sit down with me and hold my hand, or coordinate a
session for the three of us with a therapist the way some families do
nowadays. Mom simply called me up, and after some small talk about
my classes and what Aunt Inez and Aunt Carmen were up to, she
paused before casually saying, "You know, I'm getting my own apartment. I'm leaving."

I don't think Dad wanted it, though we've never really talked about
it. I certainly couldn't ask questions then. As doting as my mother
and father were, our family was from the "children are seen and not
heard" school of parenting. I may have been in college, but in their
eyes, I was still a kid who had no business offering my point of view
on grown people's problems. My opinion didn't matter, and I didn't
dare offer it. I had to grapple on my own with my feelings, which were
hard to untangle. I think I was more in shock than angry, but one
thing was certain: I hated the whole situation.

Nannie Virginia had moved to a new place, a basement apartment
in a squat, five-story building that was across the street from the
building where she'd lived since I was a baby. She was the superintendent, and she basically hooked up our entire family. My cousin
Maggie had an apartment on the third floor, and Nannie's brother
Mijo lived on the fourth. Mom got a place on the first floor, and she
also secured a studio apartment for me, where I could stay whenever
I came home.

I think Mom, cherishing her newfound freedom, got her own place
because she wanted to maintain a space of her own, but I also think
she believed she was doing something generous for me. After all, I
was on the cusp of adulthood, living independently in Binghamton.

When I came home to Manhattan from there, I could have my own apartment yet still be under the watchful eye of our family, with my mother down the hall and Nannie Virginia right downstairs. She thought it was a great setup, as did my friends.

I was already the cool kid with a car, ferrying our crew around in the Toyota Celica my dad had given me as a high school graduation present. Now my cool factor doubled, maybe even tripled, because I also had my own place in the city. My studio became an extension of my dorm room. My friends and I would make the drive downstate from Binghamton to New York on the weekends, and we didn't even bother to pack overnight bags. Instead, we kept spare outfits in the apartment to party in. We'd quickly change, then head to the Tunnel, the World, or a pop-up club. My friend Kathy still says that those were the best years of her life.

I, however, look back on them differently. I certainly had great times with my friends, who became even more like family after my parents split, keeping me anchored when I was feeling confused and adrift. But I didn't like that apartment. I'd open the kitchen cabinet and see roaches skittering by. Beyond its appearance, I hated what it represented. Home in my mind remained our place in Stuyvesant Town, which Dad held on to. But without my mother, the flowers she would place around the living room, the sound of her voice, the smell of her Chanel No. 5 perfume, it wasn't the same. Then, with Mom and I both gone, Dad took a consulting gig in Chicago and was hardly ever there. My family fell apart.

As a girl who had often been "the different one," in my old neighborhood, at school, at family reunions, having a place where I felt completely at ease was especially important to me. Now, I was coming back to an apartment that I didn't feel was mine. The family unit that had cocooned me was shattered. Everything had changed. I felt unmoored, and in many ways, I feel I've been on my own ever since, forced to grow up fast, to chart the direction of my life and career, and to just make the best of things that I was not able to change.

My extended family had always been a short walk or subway ride away, but actually being in the same building with Nannie Virginia when I was basically an adult was very difficult. She would blast me with her fiery temper. I would come home from a club only to have my grandmother throw plates at my friends, yes plates, as they walked up the stairs, yelling in Spanish for them to get the hell out.

*"Malcriada Sinvergüenza!"* she'd scream, following me into my apartment. "You're coming in with sleep on your face! You're shaming me in front of the whole block!"

I'm not sure why she cared what was on the minds of our neighbors, who included the tattooed, racist members of the Hells Angels, whose headquarters was a few doors down, especially when I'd personally seen them beat the hell out of people for no good reason. And I also didn't appreciate Nannie constantly picking up the phone to call my aunts Carmen and Inez to complain about me. "I don't know what's going on," she'd tut-tut. "She comes in with her friends at all hours of the night."

That wasn't the worst of life on Third Street. I also had a front-row seat to my mother's new life. She got a boyfriend, one who drove a motorcycle, no less. So I had to deal with this man standing where my father used to be, popping up to visit my mother, taking her out for a date on the back of his bike like she was one of my classmates. Then, at some point, he moved in. Alice in Wonderland didn't have anything on me. I felt like I had slipped through the looking glass into a world turned upside-down.

One Saturday when I was back from Binghamton for the weekend, Mom came down the hall to visit. "You act like you don't like this apartment," she said.

"I don't!" I blurted out. It was the rare moment when I decided to throw the "children should be seen and not heard" philosophy to the wind and express the hurt that I felt.

"You have your own apartment in Manhattan!" Mom said in bewilderment.

"I don't like this entire situation," I said, turning to face her. "I know I have to just deal with it, but don't ask me to act like this is a good thing and to be okay with it. And I don't like your boyfriend either. I guess you're living your fantasy life, but I'm not feeling it."

I don't recall what my mother said in response. I'm pretty sure there was silence that neither of us knew how to fill. That was so unlike me, to speak to her like that, and so out of the norm for us to be at odds with each other. My parents and I had always been so close. But during my college years, in that jittery period after the divorce, we grew apart.

I resented my father too. I didn't see him dating, but he wouldn't allow me to stay at our apartment in Stuyvesant Town without him there because he didn't want me to bring guys or anyone else over without an older adult around.

He'd always been strict, but I thought that was terribly unfair. They break up, and now I can't be in the home where I had lived from the time I was eight because he didn't want me—a college student who lived on campus—to be home by myself? In a world where I was constantly trying to find my place, I needed the familiarity of what had been my safe harbor, whether or not he was there. I felt both he and Mom were being selfish.

Now that I'm a mother, I of course understand my dad's feelings. I could have gotten into some serious mischief in that Stuyvesant Town apartment by myself. And more profoundly, the fifty-year-old woman that I am today has a completely different perspective on my mother's decision to leave my father. When I was sixteen, my mother was thirty-four years old. She was still a young, vivacious woman, finally getting to have a taste of the life that she'd sacrificed when she gave birth to me at eighteen. She was experiencing for the first time the kind of freedom and fun that I got to have every single day in college and beyond.

My parents are two of the most selfless people I know. They lived lives of deferred dreams so that their only child could be all that she envisioned. Maybe I'd been the selfish one.

At the time, I could only see how their divorce was affecting my foundation of stability, but I see things differently in hindsight and can empathize with my mother's predicament. In 1968, in our family, when you turned up pregnant at seventeen, marriage was the only option whether or not you were in love. Add to that the tumult of being part of an interracial couple in the 1960s and 1970s, having to leave your husband and daughter behind so that you could find a decent apartment, going back to school while working and raising a child—it all had to be overwhelming. Then you look up, and the daughter, who was probably the glue keeping the family together, is gone, off to college. Who wouldn't want to date, to hang out, to finally take flight and have enough breathing room to truly figure out who they wanted to be?

Adding to it all, my mother also suffered from terrible depression, which was untreated and misunderstood. She self-medicated with alcohol and later became addicted to pills. It wasn't easy.

Today, with her depression being treated and understood, my mother now says the divorce was the biggest mistake of her life, that she should have stayed married. And for me, that feeling of being lost and untethered remains fresh. It's why I have had such a hard time bringing myself to sell the current house I live in, even though we don't need as much space anymore and it's too far from where I work, from where I socialize, from the heart of things. But this large, elegant house where my children took their first steps, said their first words, and waited for Santa Claus to come down the chimney is familiar, and I never again want to feel unmoored. I knew after my parents' breakup that whenever I got married, I would be determined to stay in that union. And when I had a family of my own, I could not imagine what could ever happen to make me consider breaking it apart.

Still, a few years out of law school, I wasn't quite ready to settle down yet.

* * *

When I first moved to Baltimore to clerk for Judge Bell, he kept me busy writing the briefs that he would use to help him mold state law. But I made sure that I also had plenty of time to play, hanging out along the Inner Harbor, donning a bib and cracking crab legs seasoned with Old Bay, and making the forty-minute drive to Washington D.C. on Saturdays to stroll the endless exhibits in the Smithsonian museums.

I also dated. A lot. My boyfriends included a former state senator's son and several attorneys. One of my boyfriends was a musician, and I'd go hear him play before we'd head out to grab a late-night bite. Then there was the guy I was really crazy about. He initially planned to become a doctor, then made a hard left turn on his path to professional prestige, deciding to become a stunt man instead. My mother used to say he had the face of an angel he was so beautiful, and I was head over heels in love with him. He did go on to be quite successful as a Hollywood stunt man, but at the time we dated, I felt he lacked direction. Jumping out of windows and crashing cars to make sure movie stars stayed pretty didn't seem conducive to creating a stable environment for a family. And I needed that.

I met so many guys like that: kind, handsome, but not stable in their careers. Or they were successful professionals, but the chemistry didn't click. More specifically, they left too many boxes unchecked on my well-thought-out, quite extensive husband-to-be wish list.

At the top for me was intelligence. I'd done well in school and made a point of keeping up with what was happening in the world, so I really wanted someone who was also smart and intellectually curious. I don't care how cute a guy is; at the end of the day, we still need to be able to have a conversation. I might meet someone who was generous, interesting, funny, and then, since it was the era before email, he'd write me a letter while he was away on vacation that was full of typos or grammatical errors, and it would drive me crazy. I couldn't focus on what he was trying to say for all the mistakes that I wanted to correct with a black felt marker.

I also wanted my partner to be Catholic. That's maybe a little odd because I was never rigid about religion, never preaching or trying to impose my views on others. But it really mattered to me that my husband be of the same faith.

That may have been because when I was growing up, my dad tried a lot of different religions. He was born-again one minute, attending a talking-in-tongues, Holy Rolling storefront church, then a few weeks later he'd be immersed in a Buddhist temple and chanting in front of an altar he set up at home. Mom, on the other hand, went to the same church, Nativity, her entire life. Because Dad was always experimenting, we rarely worshipped together as a family. I would just attend Mass with Mom, and that never seemed right to me. Once my parents divorced, I think I sometimes wondered if we'd shared the same religious practice, might those rituals have kept our family together? They probably wouldn't have, but I felt that any little thing that could fortify a marriage, that might keep it from disintegrating, was worth pursuing, so I felt that sharing a spiritual connection with whomever became my husband couldn't hurt.

My list didn't stop there. He had to be loyal. I wanted someone with a serious profession who could provide financial stability. And I wanted him to speak Spanish, or to be bilingual, whatever his second language was. When I began traveling, it always struck me as ridiculous that the United States was pretty much the only country in the world where people think speaking a single language is fine while virtually everywhere else, people are able to communicate in a myriad of ways. I romanticized the idea of raising children with a multilingual world traveler, in a home teeming with culture.

Oh, and he had to be over six feet tall. Had to be. I was five foot six, taller than the average girl, and I wanted someone I had to rise up on my tippy toes to kiss.

Needless to say, finding someone who met my extensive criteria wasn't easy, and after a while I grew frustrated from meeting men, having fun for a short while, and then realizing that he was definitely not my forever guy. I wanted children. I didn't want to still be dating when I was on the verge of retiring, and my mother was terrified she would never be a grandmother.

"You know what your problem is?" Mom said after hearing again that I didn't have a steady boyfriend. "You don't go to church enough."

I don't know what solace and advice I was looking for. But that wasn't it.

"Yeah, that's my problem," I said, my voice thick with sarcasm. "I'm going to meet the man of my dreams in church." I wasn't going to church often, but the past few times I'd been, I could count the number of men on two hands, including the priest. And if the male parishioners weren't already married, they were a whole lot older than me.

"Do you have a home church? You've been in Baltimore almost a year and you don't have a home church," Mom continued accusingly.

"Yeah, yeah, Mom," I said, now itching to get off the phone. "I've got to find a home church. That will solve all my problems."

Even though my dating life was less than satisfactory, I had plenty to keep me busy. I'd jogged when I was in college, and I'd taken it up again after moving to Maryland. I was entering 5K races to prepare for an upcoming marathon I wanted to compete in.

One morning, after running a couple of miles, I was almost home. St. Philip and James Church, an ornate sanctuary a block and a half from my apartment on Charles Street, was on my right.

I'd walked by St. Philip's many times but never gone in, and I certainly wasn't dressed to attend service that day. My hair was slicked back in a ponytail, and the sweat suit I was wearing was living up to its name. But that mischievous streak that had gotten me into trouble so often at St. Anselm and Dominican was still a part of me. Since Mom believed I'd find my soul mate once I got back into church, I figured I'd drop in, then call to tell her I'd followed her advice and indeed found the man of my dreams. Not!

It was the noon Mass, and I found a seat in a back pew. Funny, I thought. The church was full of people my age. I wasn't expecting that, though I later realized that they were students from Johns Hopkins, which was practically across the street.

I'd been there about five minutes when I turned around. And in walked Emmanuel "Manny" Hostin.

He was gorgeous, decked out in a beautifully tailored suit, with curly dark hair that framed his deep-set eyes. He was with a friend, and the two of them grabbed a seat a few pews in front of me. From then on, I wasn't focused on the Nicene Creed or the Penitential Rite. I may have absentmindedly hummed along with the choir, but I didn't hear a single word the soloist sang. All I could think about was how I could meet this man. I couldn't go up to receive communion dressed like I was. I looked ridiculous. I would have to sit and wait. My eyes were glued to his every move.

When the service finally came to an end and he and his friend got up to leave, I followed them outside. He went into Sam's, a bagel shop around the corner. I got in line, right behind him.

"Hello," I chirped. "How are you?"

"Hello," Manny said. His eyes didn't exactly light up.

"Wasn't that a lovely service?" I piped up again.

He looked a little surprised. "You mean at St. Philip and James?"

"Yes! Wasn't it an insightful homily?"

He looked me up and down, from the rubber band holding my ponytail in place to my athletic socks and sneakers. "You went to church dressed like that?"

Knowing Manny as well as I do now, I know that it was bordering on blasphemous in his view to go to church dressed in jeans, let alone sweatpants. But I ignored the dig. I'd come too far to back off now. I explained that I was training for a marathon.

"Oh," he added. "I used to run when I was an undergrad." When I asked where he'd gone, he told me Johns Hopkins, so I thought he might be from Baltimore. But he told me no, he was from New York.

New York? Besides being my hometown, where I hoped to eventually move back to, New York roots were also a sign that he was a cosmopolitan guy. My list began to unspool in my head.

What was he studying?

"I'm in medical school."

*Check!*

What specialty was he pursuing? He was going to be an orthopedic surgeon.

*Check, check!*

He looked a little Middle Eastern. "Are you from New York originally?" I asked casually. No, he replied. He was born in Spain to a Haitian father and Spanish mother.

*"Habla español?"* I asked, my heart starting to pump so hard I thought it might leap out of my chest.

*"Seguro que sí,"* he said.

"Oh my freaking God," I thought to myself. "My mother tells me to go to church, and this guy literally has everything that I freaking want!" The joke was on me, not Mom. And I didn't mind one bit.

We got to the counter, and Manny turned away to order his sesame seed bagel. He'd answered my twenty questions, but it was obvious he wasn't checking for me. His friend, on the other hand, who was a lot more talkative and friendly, mentioned that he and Manny liked to throw parties. When he offered his phone number, I took it.

Forget jogging. I sprinted back to my apartment building, raced upstairs, and called my friend Lisa.

"Lisa!" I yelled into the phone. "I met the man I'm going to marry. But he doesn't know it yet."

She knew my list and was as excited as I was.

"Does he speak Spanish?" Lisa asked.

"Yep!"

"What does he do?"

"He's in medical school at Johns Hopkins. He is going to be an orthopedic surgeon!" I screamed.

"Is he over six feet tall?" she asked, her voice jumping up an octave.

"Uhhh . . ."

The truth was that my husband-to-be was short and I hadn't even noticed. He walked so tall. But now thinking about it, I could look him straight in the eye and would have to bend my head to kiss him when I put on heels, short. I told Lisa, "I think he's my height."

As fate would have it, I ran into Manny's friend at a party a week or two later, and this time I wasn't covered in perspiration or wearing a sweat suit. My hair was very long at the time, way past my shoulders, and I was wearing it loose. My eyeshadow and lipstick were on point, and I was wearing earrings and a cute, tight little burgundy dress. He spotted me first and came over to where I was standing. He and Manny were going to be throwing one of those parties he'd told me about when we met. He said that I should stop by.

Manny later told me that his roommate came home that night and told him that he'd run into the girl they'd seen at the bagel shop— and she was seriously hot.

"What girl? The one in the polyester sweat suit?" Manny asked dismissively. It wasn't actually polyester. It was cotton and emblazoned with the name of my sorority. But those details didn't matter to Manny, who always wore a jacket and tie to service. "I don't believe it."

"No, man. She is beautiful! Trust me, dude."

When I made my way to their apartment a few days later, Manny opened the door. He didn't recognize me, but I said it was nice to see him again. When he asked if we had met before, I told him yes, we'd met in church. That's still our running joke. He would tell folks that we'd met in a bagel shop, but I'd let him and everyone else know that we'd met in church. That's where the connection happened. He just hadn't seen me.

Manny actually had a girlfriend at the time, but he broke up with her not long after that party. Two years later, we got married, and I became Asunción Cummings Hostin.

We had a huge wedding at the same mansion where the actor Will Smith married Jada Pinkett. There was a martini bar, common at ceremonies today but more of a novelty then, with cute, sort of corny signs that we made up, with sayings like "Drink two of these, but don't call Dr. Hostin in the morning!" And we had a band that played salsa, then a DJ to remix old-school jams by New Edition and the Notorious B.I.G. later in the evening.

For one of the few times in my life, those who peopled my many worlds were all gathered in one place. Nannie Mary and Nannie Virginia, Aunt Carmen, Uncle Joey, my cousin Jeff, Sean, and all my down-South family partied alongside Judge Bell, Jim Gillece, Dando, and many of my fellow prosecutors in the Justice Department. Lisa, my co-clerk who'd first heard my prediction that I'd met my love match, gave a toast, along with Kathy, Regina, Tamika, Stephanie, and my crew. Folks whose paths might never have crossed beyond the mansion's door were having a ball, talking, celebrating, and honestly getting pretty lit off those martinis, all together.

The reception got going at noon and was supposed to end about three, but when the band started preparing to pack up, my father, ecstatic and buzzed as much from pure joy as he was the champagne and cocktails, ran over to ask how much it would cost to keep the party going a few more hours.

"Probably a lot," I told him, exhilarated but exhausted.

"Done!" Dad said, spinning around and heading back to the dance floor. The next day, when I spoke to my father again, I reminded him that he'd volunteered to spend more money on a wedding that had already been quite expensive. He paused for a second, then said, "You're my only child. For you, only the best! Besides, you only get married once."

He was right about that. And we had a wonderful time.

* * *

Manny and I had been married about two years when we started talking about having a family. *Time* and *Newsweek* magazines seemed to have an article in every other issue warning how once you turned thirty-five, your eggs turned a corner and it would become more and more difficult to conceive. I was thirty-three and dreamed of having three, maybe four kids, so we started trying. But nothing was happening.

It's one of life's many ironies. You worry about getting pregnant before you're ready, a soundtrack that definitely played in my head given that my own parents were teenagers when I was born, but now that I was eager and ready for children, it was proving to be a struggle. We never used birth control during our marriage, yet I never got pregnant.

My primary care doctor sent me to a fertility specialist, an amazing physician named Fady Sharara, and we discovered that I had fibroids, benign uterine tumors that are particularly common in women of color. Dr. Sharara told me that there were several, of varying sizes, with some so big that my uterus had stretched to the size of that of a woman who was five months pregnant.

The fibroids on their own weren't enough to prevent me from carrying a baby. But when Dr. Sharara tested Manny, we found out that Manny's sperm were so overactive, they were literally knocking one another out. It was a double whammy that was going to make it really hard to conceive. Dr. Sharara said I needed to think about having an operation, known as a myomectomy, to remove the fibroids.

I was terrified. I'd never had surgery before, not to take out my appendix, not to remove my tonsils, not to fix a broken bone. And I'm very risk averse. Though I've traveled extensively, I'm usually anxious getting to where I'm going because I don't like to fly. I will zoom around in a go-kart at an amusement park, but I don't ride roller coasters. And because I don't feel like chasing away nightmares once I go to sleep, I avoid scary movies.

So the thought of being wheeled into an operating room put me into a full-fledged panic. We tried a procedure that involved "washing" Manny's sperm, to slow them down and increase the chance of my becoming pregnant. I did, twice. But my uterus was so misshapen by the fibroids, I miscarried both times. There was no choice. I had to have the surgery.

The day they gave me the anesthesia and prepared me for the operation, I started crying. On the way to the operating room, I even asked Dr. Sharara if I was going to die. I was afraid I would drift off to sleep and never wake up. But I wanted a baby so much, it was worth taking that chance. It was worth anything. It was worth everything.

Dr. Sharara later told me that it was one of the most difficult surgeries he'd ever done. I had twenty-six fibroids, the most he'd ever had to remove. He actually took a picture of them that I still have.

Though the surgery was successful, I was still likely to have a high-risk pregnancy given the condition of my uterus. Manny and I briefly considered using a surrogate. Years later, I talked to the actress Gabrielle Union on *The View* about surrogacy, since she and her husband, the basketball player Dwyane Wade, went that route to have their baby girl. But neither Manny nor I were very comfortable with the idea. I really wanted to try to carry my own child, if there was any way I could.

We decided to try in vitro fertilization. I had pretty good health insurance working for the Justice Department, enough to cover one, maybe two cycles. Each would cost $30,000.

There was no way to anticipate the emotional vortex we were about to enter. Each cycle takes weeks. I was injecting myself with medications like Lupron, to produce more than the single egg women develop in a typical, natural cycle. My doctors checked my blood repeatedly to see how the eggs were developing, and then before I ovulated, I had to undergo a surgical procedure to retrieve the eggs.

After that, the eggs were placed in a dish with Manny's sperm. If an egg was fertilized, an embryo developed, and it would be implanted inside my uterus. Again, there would be more monitoring as the fertility specialists checked for the amount of human chorionic gonadotropin I had in my blood to determine if I was indeed pregnant.

Your anxiety grows with every step, but so does your hope. And then, if the day comes that the hormone level drops, telling you the embryo didn't attach or survive after all, the cycle starts all over again.

I got pregnant during the first round. Then, I miscarried.

Manny's parents, my mom and dad, all my girlfriends, had been beyond excited. And then I had to tell everybody that the pregnancy was over, just like that. For someone who hated roller coasters, emotionally I was on the tallest, steepest, most frightening one I had ever encountered.

When the insurance would no longer pay for the treatments, Manny and I started paying out of pocket. We had some savings, and we'd bought a house in Baltimore. We took out a home equity line of credit to help cover the costs.

I went through another cycle, and once again we found out we were having a baby. But there was miscarriage after miscarriage after miscarriage. Including the two I'd had before we began in vitro, I'd now miscarried five times. I was able to get pregnant, but I couldn't carry anywhere close to term. A pregnancy might last three weeks, or it might last a month. Once I made it to six weeks. But no matter how long or short the pregnancy was, it was always devastating when it ended, because of all we'd done to get to that point and all the hope we'd held on to, believing that this time it would work out.

The hormones contained in the injections I had to give myself made me insane, like I was experiencing the emotional highs and lows of PMS multiplied by ten. And I was giving myself shots, making regular visits to the doctor, and reeling from miscarriage after miscarriage, all while still working full-time as a federal prosecutor. Manny was in the early stages of a fellowship at the University of Pennsylvania in Philadelphia, commuting home on weekends, and our marriage was still young. It was an incredibly difficult chapter in our lives.

\* \* \*

I didn't share what I was doing with many beyond our immediate circle of family and close friends, but my exhaustion and moodiness were visible to anyone who cared to notice. Most were kind, but there was one notable exception, a deputy chief in the U.S. Attorney's office. She was another black woman I'd initially assumed would be an ally since we were both African American women working in a competitive, high-pressure field. But she never liked me. She was always wanting updates on whatever I was working on, like I was an intern who constantly slacked off rather than a colleague who would stay up all night prepping for trial if I had to.

I would drag myself out of bed when my alarm clock sounded off at six in the morning. Even without my explaining the grueling medical process I was going through, the detectives I worked with were understanding when I told them that getting to the office by 7:30 A.M. to go over a case was taking a toll on me. They agreed to just meet me downtown at the courthouse at 8:40 for a brief catch-up before we headed to court.

I was popular with the security guards at the courthouse as well, and they made other concessions, like letting me park in front of the building instead of my having to grab a spot blocks away, then racing, my leather bag bursting with files and briefs, to be in front of a judge on time.

But my deputy chief didn't cut me an ounce of slack. Not one. My administrative assistant told me that she would check to see if I was in my office every morning. I noticed she would constantly ask my paralegal for information on what I was doing on my cases. She would also ask agents and detectives about my case preparation. Then, she wanted me to give her a weekly status report about my cases. She didn't ask this of the other prosecutors. The day came when I'd had enough.

I'd slipped off my pumps, propped my feet up on my desk, and was reading the *Washington Post* when she walked in. I don't think she even knocked.

"I would like to speak to you," she said.

"Okay." I kept on reading. My feet didn't budge. My Stuart Weitzman heels remained perched by the trash can.

"I said," her voice rising, "I need to speak with you."

I peered over the top of the paper. "When. I'm. Ready."

Then I let loose. "I don't appreciate you treating me like an intern when I have one of the best track records in the trial unit! It's not right, and I'm starting to think it's petty and personal. I'm tired of it!"

I don't suffer fools, but I'm always professional. I can cut you, and you won't know what hit you until you're at home hours later, having a glass of wine, reflecting on your day, and suddenly realize that you're bleeding. But after all those doctor visits, all that hoping that I was finally pregnant only to find out repeatedly in the morning that I was not, pushed me over the edge. I couldn't hold back. The South Bronx swooped in from around the way, bubbled over, and smacked her.

She stormed out. Then she wrote me up, recommending that I receive a formal reprimand for being insubordinate and doing substandard work.

When the head of the unit called me in to find out what was going on, I finally had to tell someone beyond my closest relatives and friends about the IVF. I explained that I was trying to get pregnant, taking hormones that made my moods pivot like a seesaw, and I felt that the deputy chief was on me in a way that was unnecessary. What I'd tolerated before felt unbearable now.

Our unit head made the deputy chief change her complaint to say that I'd behaved out of character because of a medical issue I was

dealing with, so the incident never escalated to a more formal punishment. But the whole situation hurt, particularly because of all people,
it was another black woman who wanted to put something in my file
that could have followed me throughout my career.

I shouldn't have lashed out at her. I should have responded to her
request to talk right when she made it. Still, I felt she had a pattern of
behavior toward me that bordered on hostile. I worked hard, tried to
be cordial, and it made no difference.

She may have resented my popularity with the detectives and others in the office, though that was no reason to mistreat me. Our jobs
were hard enough, and while it's absurd to think every single black
person in the world is going to be friends, I do believe that we should
show some solidarity or at the very least not undermine one another,
given all the challenges we face.

* * *

My body had been through a lot. My fibroid surgery had left my
uterus full of scar tissue. And after round after round of treatments
and so many miscarriages, I no longer wanted to tell anyone when I
got pregnant. The sad looks that they'd try to quickly erase, and their
hurried phrases to try to cheer me up, just made the disappointment
more painful.

Finally, after what felt like the umpteenth cycle, we had six embryos. Usually, a woman will have one or two embryos implanted
at the most. But Dr. Sharara said that he wanted to put in all six to
increase the chances that I could bring at least one to term.

That seemed crazy to me. What if I got pregnant with triplets?
I thought. How could my fragile uterus withstand that? And while

some women may have chosen to "selectively reduce," winnowing the number of embryos once they were in utero, that wasn't an option for me and Manny. We were Catholic, and that would be akin to abortion, violating our religious beliefs. So if all the embryos were viable, we could wind up with several babies all at once. By those same religious principles, if my life became endangered, we would need to save the babies no matter what. Anticipating all that could happen, all that could go wrong, was excruciating.

Even with all the embryos, it was far from a sure thing that even one or two would survive once implanted. Dr. Sharara said that we just had to hope at least one was strong enough to find a spot to latch on to in my womb and thrive.

Our money had run out, and so, perhaps, had time. Manny and I decided to go for it. But we told no one, not even our parents. The odds were long. The odds were against us.

When I went for a follow-up visit, the pregnancy hormone was very high, which is usually the case when there is more than one embryo. But within two weeks it became clear that only one embryo had survived. The baby had found a place to grow. It was our first miracle.

My hope began to grow, like the baby inside me. After six weeks, I asked Dr. Sharara if it was finally safe to tell people. "Is it happening?" I asked him tentatively. "It's really happening," he said reassuringly.

We told our closest family members and friends. What was even more wonderful was that Kathy, Regina, and Tamika, my three best friends, were pregnant at the same time. We were going to be able to shop for baby clothes, decorate our nurseries, and chart our cravings together.

Despite the fibroid surgery, I was very healthy overall. I was only thirty-three, and the baby's heartbeat was strong. My doctors marveled at how out of six implanted embryos, one had made it, negating the need to ponder how I would care for several babies born at once. These were all great signs, they said, so there was no need to be nervous.

I was still heading to the courthouse every day, toting my heavy litigation bag, while Manny was completing his fellowship in Philadelphia. We were also planning to ultimately settle back in New York so we'd be closer to family when the baby was born. We sold our house and found a comfortable apartment where we could stay until we made the move north. Occasionally, on the weekend, I drove or took the train to Westchester County to house hunt. It was about an hour outside Manhattan, but far enough away that our budget could stretch further, enabling us to get a larger home with maybe a bit of land. But I was doing too much.

When I passed the three-month mark, the point in time when the riskiest period of a pregnancy is usually considered over, I began to tell people at work that Manny and I were expecting a baby. In my head, I was preparing to leave the U.S. Attorney's office in Washington D.C. to possibly transfer to its counterpart in the Southern District of New York. I had made dear friends in Maryland and Virginia and learned so much during my time working there, but I couldn't wait to move back home.

When I got back to Maryland after another house-hunting trip, I put down my bag in the living room, flipped on the television, and settled in a chair to watch *Oprah*. All of a sudden, I felt something wet between my legs, then a gush of blood.

"Oh my God!" I screamed. Not again. My pants were completely soaked with bright red blood.

I became hysterical. I grabbed the phone and dialed Manny, who was an hour and a half away in Philadelphia. He told me to call Pierre, his best friend, who was an emergency room physician. Scared to death and not thinking straight, I said that I couldn't call Pierre. He was a man. I was bleeding between my legs. I couldn't see him. He wouldn't understand.

"Call Pierre," Manny said, trying to calm me down. "He's going to come and get you to take you to the hospital."

But I couldn't do it. I had lunged for the phone to reach Manny, but now I didn't want to move; I didn't want to walk. After so many miscarriages, I couldn't bear one more. And if I wasn't still, I feared it would surely happen again, if I hadn't lost the baby already.

Manny hung up with me and called Pierre himself. He lived only a few minutes away. About fifteen minutes later, Pierre was banging on my front door. I still didn't want to move, but the thought that I had to do something if there was a chance to save my baby snapped me out of my paralysis. Hunched over, I took a few steps toward the bathroom, yanked a towel off the rack, then hobbled over to let Pierre in.

Pierre wanted me to let him check me, but I didn't want him to. I just gave him a glimpse of my legs, streaked with blood. He called the emergency room at Howard County General, where he was doing an emergency-medicine rotation, and then carried me down a flight of stairs into his car.

It's bizarre the details that stick in your mind from a moment when you were nearly suffocated by fear. I remember Pierre had a silver BMW convertible, shiny and pristine. He had thought to quickly grab a few more towels from the apartment, which he laid out on the passenger seat, but I kept thinking they weren't going to do any good.

I was hemorrhaging, and I was horrified that I was going to bleed all over his car.

Other thoughts crystallized later, like the cruel realization that status and class can be the difference between life and death when others look at you and cannot see past your race. Howard County was a public hospital, with an ER room full of mostly uninsured in pain. Like hospitals across this country, where health care is still viewed by too many as a privilege rather than as a right, the personnel admitting patients seemed more preoccupied with seeing incoming patients' insurance cards then getting them in immediately to see a doctor. Pierre wasn't going to let me linger in the ER, where the hospital staff wanted me to wait.

"This is Dr. Hostin's wife, and I'm Dr. Vigilance," he said firmly in his clipped British accent. "I have privileges to practice here. She's thirteen weeks pregnant. She's a federal prosecutor. You're going to admit her. And I don't want a wheelchair. Bring out a stretcher."

The hospital employees snapped to attention, and things started moving very quickly. But there were more hurdles to come.

After I was admitted, the nurse who was dealing with me said that they were going to call in a general practitioner for the examination. Pierre had to insist that I see someone from the obstetrics and gynecology department. Why did he have to demand that an obstetrician see a pregnant woman who was bleeding profusely? It was painful to see the type of medical care that people of color, and particularly black women, often receive, their health challenges ignored or dismissed as less important.

A *New York Times* article in 2018 reported that black babies in the US were more than twice as likely to die as those who were white, and black women were three to four times more likely than their white peers to lose their lives because of complications connected to pregnancy or childbirth. Racial bias plays a tragically large role in those outcomes. Even tennis superstar Serena Williams, who suffered a pulmonary embolism after giving birth to her daughter, said in *Vogue* magazine that her complaints and symptoms were initially ignored. Wealth and celebrity were not enough to insulate a black woman from the callousness of systems and institutions that still, too often, see us as less worthy.

Pierre, a black man, knew what questions to ask, and as a doctor with privileges at that hospital, he had the clout and the confidence to demand that I be treated with care. But a pregnant woman who was hemorrhaging shouldn't have needed an advocate to make sure that she was put on a stretcher, that she was able to see an ob-gyn, that she was treated with compassion and respect.

When I finally saw a specialist, I learned that my placenta had ripped. The doctor said that it was one of the worst tears he'd ever seen. Luckily, the baby's umbilical cord was attached to the part of my placenta that was healthy and still intact. A short time later Manny arrived. To get there so quickly from Philadelphia, he must have dropped whatever he was doing and driven over one hundred miles an hour. He ran in, panting, still wearing his scrubs.

Even in the presence of Manny and Pierre, two doctors, one of whom was my husband, the callous behavior that had greeted me at the hospital door continued. We were told that given the tear and the already weak condition of my uterus, there was a one in four chance the pregnancy would go full term.

"What do I need to do now?" I asked, tired and still frightened.

"Well," the doctor said matter-of-factly, "make a follow-up appointment with your regular ob-gyn."

That was it. He had just told me my baby had a 25 percent chance of surviving, I'd suffered the worst placental tear he'd ever seen, and he was just going to release me, with no directions, no advice.

At that moment, I looked at the sonogram, and I swear, on the monitor, I could see my son's tiny hand, and it was waving at me. It was almost as if he was saying, "I'm good."

Seeing that gave me strength. This doctor's attitude was unacceptable. I had to fight. The prosecutor, the girl from the streets of the South Bronx, and the mother-to-be all came pouring out.

"Excuse me," I said, steel in my voice. "What do I need to do to make sure that this baby makes it? You're looking at the sonogram. He's fine now. What do I do?"

"Well," the doctor said, still more nonchalant than he should have been, "you'd have to be on strict bed rest for the rest of your pregnancy. That means no movement. And you would have to wear a fetal monitor. And again, no movement."

So that's what I did.

* * *

Manny and Pierre took me back home, and I called my mother. She immediately asked for and got a leave of absence from her job, though I know she would have quit if the school where she was working had said no. We got a bedpan and rented a fetal monitor from the hospital, and within a couple of days, Mom and her two poodles had moved in to the apartment.

My friends came over frequently to keep me company as I sat propped up in bed. And the U.S. Attorney's office was amazing. They allowed me to transfer from the trial division, where I'd been working, to the appellate division, which didn't require me to go to court. I whiled away the time writing briefs. And the office came through in another critical moment. Like so many workers trying to make do with our country's crippled system, the Justice Department didn't offer much sick time. I'd been on bed rest for just a couple of weeks when I got a call from the human resources department, informing me that I was about to run out of sick days, meaning that I would no longer be paid while I was home trying to keep my baby alive.

When I told the head of the unit, he sent out an email that told my colleagues what I was going through and asking, if they were willing, to donate some of their personal days to a leave bank being set up for me. So many prosecutors donated that I was covered for the next six months, the remaining duration of my pregnancy. Out of all the presents I would receive, that was no doubt the most thoughtful.

I watched *Maury Povich* episodes on a loop—you *are the* father! I needed those distractions to ease my anxiety, but I could relax for only so long. When you're on bed rest, forbidden to move, you lose all dignity. My mother gave me sponge baths, and then when Manny, my relatively new husband, came home, he would give her a break, emptying my bedpan and literally wiping my backside. My now plus-fifty-pounds backside.

The doctors had spun out a bunch of statistics that once again showed the peril that shadows black boys even before they're born. In at-risk pregnancies, they were the least likely to make it, though because my son was biracial, the physicians said the equation might be a bit different. Still, they said, if he was born prematurely, a higher fetal weight could improve the baby's chances to survive.

So Mom was on a mission to fatten me up. I'm talking heaping plates of beans and rice, *chicharrones,* and stewed chicken served around the clock. The smoke detector got a real workout from all the sizzling grease crackling in my kitchen.

I had always been small, no bigger than a size 2 or 4, but lying in that bed, eating all day, I gained seventy pounds. Manny watched the batches of food flowing in and out of our bedroom and said he'd never seen anything like it. But I think what he really meant, yet was too kind to say, was that he'd never seen anything like me. I resembled a beached whale. I would leave the bed only once a month, to go to my doctor's office, and there I would be weighed. That was the worst, getting on that scale. I watched the nurse move the weight farther and farther to the right to balance it. "Oh my God," I thought. "I've never seen those numbers before."

It was by far the heaviest I'd ever been. I fell into a deep depression. Given my mother's history of depression, I was scared. I also began reading countless blogs about fertility, full of passages about torn placentas, IVF treatments, and the odds of prematurely born babies surviving. It was like sliding down a rabbit hole. Some of the information was helpful, reminding me that I wasn't alone and that many women had come through this ordeal and had a joyful ending. But it was overwhelming.

If my mother was worried, she didn't show it. She couldn't wait for the birth of her first grandchild. Mom must have spent thousands of dollars buying onesies and beanies, blankets and shoes. When they arrived in the mail or she returned from a short shopping trip, she couldn't wait to show them to me. But I didn't want to see them.

I still wasn't convinced that the baby would make it. I became obsessed with his heartbeat. We eventually returned the hospital fetal monitor, which cost a fortune to rent, but we were able to lease a similar, slightly cheaper device. I would put the cold ultrasound gel on my abdomen probably every hour. If I couldn't immediately find the baby's heartbeat, I was like a maniac, convinced that I would never hear it again. I came to know that if I drank orange juice, his heart would speed up; if I ate too much heavy greasy food, it seemed to go down. I was going out of my mind.

One day, Mom held her latest baby blue purchase up to the light, and I snapped. "Stop buying all this stuff! I don't want to see it," I screamed.

That was it. She made her way past the boxes piling up in the living room and picked up the phone. She was calling in the big guns: Sister Eileen.

Sister Eileen had been a fixture in my life when I was younger, occasionally accompanying her brother, Father Francis, to our family dinners at Nannie Virginia's. When I was young, she was always available to chat, whether I had a question about life here on earth or in the hereafter. I hadn't seen her in years, but Mom had maintained their connection. I'm so thankful they did.

Sister Eileen began to call me every day. There had always been something about her voice, her spirit, that calmed me. And now when I needed her most, she once again soothed my nerves.

"We are going to pray this baby into being," she told me, as I quietly cried. "You deserve this blessing."

The next day, the phone would ring again. "He is meant to be," she said. "He's going to be here. Mother Mary will protect him his whole life, and you will be a great mother."

She told me to feel the wonder of this miraculous experience. "You're going to look at those baby clothes your mother bought," she said. "Cherish these moments."

With her words, I was lifted. It is a shame that the church doesn't allow women to be priests, because Sister Eileen could have been a great one. The ministering she gave me, with just a few minutes each day on the phone, helped me to finally see light in the midst of a period that should have been joyful from the beginning but until then had been mostly bleak and dark.

Reluctantly, I started to look at the clothes. Mom had bought enough that my baby boy could probably wear a different outfit every day for his entire first year. I was still nervous, but I no longer spent every minute either agonizing about my baby's safety or fretting about my weight. I started getting ready mentally to be a mother.

In my thirty-second week, I made my monthly trip to the doctor. Manny as always made a point of being back in Maryland to go with me.

The doctor checked my sonogram, put a stethoscope on my belly, then uttered the sweetest words I'd ever heard. "If he's born today, he'll be fine," he said. "He's big enough, and you're close enough to full term."

I was stunned. I had been on my back twenty-four hours a day for nineteen weeks.

"I can walk around? I don't have to stay in bed?" I asked, disbelieving.

He warned me to still take it easy. But no, I didn't have to stay confined to my bed anymore.

I looked at Manny. "We did it," I said, tears in my eyes.

"You," he said. "You did it."

The pregnancy, which had seemed endless, interminable before, now seemed to fly by. My mother and friends threw me a huge baby shower. A lot of the prosecutors I'd worked with came. And I actually enjoyed it.

\* \* \*

My favorite day of the pregnancy came a week or so later. It was July, the height of summer, when on the sunniest days, the humidity hung like raindrops.

My paranoia about the pregnancy had eased, but my massive weight gain continued to embarrass me, and carrying all those extra pounds in the heat made me miserable. Manny suggested we go down to the pool in our building.

"No! I'm so fat!" I protested, but he insisted.

I didn't have a bathing suit that could fit me, but I put on shorts and a T-shirt, and we went downstairs. Manny guided me gently into the water and turned me on my back. Then, supporting me with his hands, he carried me. I felt weightless floating around that beautiful pool. Manny literally took the weight of the past nine months off me. Manny was, and is, a remarkable man. It was the most wonderful day.

"You did it," he said softly.

"Yes," I said. "I can't believe it."

On August 15, 2002, through a cesarean delivery, I gave birth to my son. All that rice and beans paid off, because he was over eight pounds, and very long. Manny and I joke with him now that he was the $60,000 baby, but of course he was worth it. He was worth everything. Our entire families came to the hospital to welcome Gabriel Cummings Hostin into the world. My father exclaimed that he was clearly the smartest baby in the nursery. Nannie Virginia held him and exclaimed that he looked just like her father.

Pierre, who picked me up the day that my placenta wall tore and rushed me to the hospital, is his godfather. My son is alive because of Pierre. I know this to be true. And to this day I tell Pierre so.

We named our baby Gabriel, because like God's messenger in the Bible, he was an angel.

We played around with other names. Manny, named for his father, Emmanuel, rejected the idea of having a Junior because he said he wanted our son to be his own person, but we briefly pondered Emilio. Yet we kept coming back to Gabriel. It felt right. It felt perfect.

Gabriel, even as a teenager, is so loving and kind, never getting off the phone without telling me how he feels. "I miss you, Mama," he says when he is away on a trip. "I love you, Mama." And he has always been the strongest kid, just fearless, with a wanderlust that took him to Australia, Japan, the Galápagos Islands, and the forests of Ecuador to do community service projects before he was out of high school.

He also loves the water. He was scuba certified at the age of twelve. I think Gabriel was meant to enjoy the world because it was so hard for him to get here. He is so kind and compassionate because he was the kid who wasn't supposed to make it.

They say that mothers, so overtaken by the emotions that come with having a child, forget the pain of labor as soon as they give birth. I think there's some truth to that, because even after the ordeal I went through to have Gabriel, I wanted more children. I underwent IVF again.

The first cycle was unsuccessful, but the second time, we had two embryos. For two to three months, we were expecting twins, which filled me with joy since I wanted three or four children. But the pregnancy hormone dropped, and I learned that then, there was just one.

Still, in contrast to my previous pregnancy, I was actually able to enjoy this one more. I was on modified bed rest because some fibroids had come back and I didn't want any more of those surgeries, but I could still walk around and be somewhat active. I did have one scare, when I experienced some bleeding and again was rushed to the hospital. But the baby was fine and strong. I gave birth to my beautiful seven-pound baby girl on May 4, 2006. Paloma Cummings Hostin is fierce and independent. She is the girl I dreamed of. My mini me in so many ways. A joy.

I loved being a mother, so much so that I was actually open to undergoing IVF yet again to have more children. In the end, Manny was the one who said no. Though I had an easier time with Paloma, he couldn't shake the memory of what had gone on with Gabriel. He said that he didn't want to see me become that person again, the woman who had been so depressed, so frightened, so paranoid. When people talk about IVF, the focus is usually on the woman and her struggles. But fathers suffer greatly too.

I have the deepest of connections with my children, whom I would give my life for. But the truth is that the journey through IVF can be soul crushing. I never again want to be in the place that I was at two in the morning, three months pregnant, lying in bed with a fetal monitor, alone. I never want to revisit all those moments spent thinking that my baby was going to die inside me at any minute. After three failed IVF cycles, it was a terrifying place to be, even for the strongest, most hopeful, most religious person, even for someone in a happy marriage, with a mother who upended her whole life to be there with her, surrounded by a supportive family, deeply loving friends—and one amazing nun.

In the years after Gabriel's birth, and that of my daughter, Paloma, I again lost touch with Sister Eileen. She'd retired and moved away from New York City. But a couple of years ago, she was on my mind. I asked a good friend, Father Edward Beck, a contributor to CNN, if he could help me find her. When he did, I discovered that she was in the hospital, dying from cancer.

This time, *I tried to give her comfort.* I told her how important she had been in my life, what it had meant to have her pray light into the darkness when I was paralyzed with worry and fear. I told her that she had prayed my sweet son, Gabriel, into existence and that he was more wonderful than I could have ever dreamed.

She died not long after. But I'd found her while there was still time to say thank you. And I did.

# CHAPTER EIGHT

# THE DREAM DEFERRED

B y the time I'd outgrown *Sesame Street* and *The Flintstones* in the early 1970s, it was no longer rare to see African Americans on the small screen. But I still felt a ripple of excitement when I flicked on the TV and saw that black people like me were there.

Though books were the main source of entertainment in our home, Mom, Dad, and I watched *60 Minutes,* the iconic news show, every Sunday night, and we took particular pride when Ed Bradley, the debonair African American journalist known for his sharp inquiries and rich voice, reported one of its segments.

As a kid, I would sing along to the theme of *The Jeffersons,* a sitcom about a black family that ran a successful business and moved to an Upper East Side high-rise. And when I was at SUNY Binghamton, I'd skip out on a game of spades in the Union to rush home and watch the latest life lesson preached by Clair and Heathcliff Huxtable on *The Cosby Show.*

I recognized the power of images, and when I began studying communications, my desire to help shape them grew.

I enjoyed writing, but I really felt alive when I stitched together pieces in the SUNY communications department's edit bay. One of my class projects was to put together a video promoting the Equal Opportunity Program, which gave me financial aid and provided assistance and tutoring to other students of color, many of whom were the first in their families to go to college. It was fascinating learning how to tell a story through video, which in the 1980s was still a relatively cutting-edge medium as TV transitioned from film.

I leaned in to broadcast even more when I got a gig at the campus radio station, WHRW, which could be heard throughout the county. My partner was a guy named Fritz, and with the prime-time slots already claimed by older students, we took over the midnight to 2:00 A.M. show, bantering back and forth in between spinning love songs by Luther Vandross and James Ingram. By my senior year, I was heading to the studio and broadcasting three mornings a week.

Even though my mother later made it clear that a future for me working in media terrified her, she did help me land an internship with the broadcasting trailblazer Carol Jenkins. Jenkins was one of the first African American women to be a television anchor in New York City, starting at WNBC in the early 1970s. Mom had taught her daughter in elementary school, and she asked Carol if I could shadow her on the job. Carol was gracious enough to welcome me, and I trailed her as she prepared for on-air interviews with local politicians and covered fires, scandals, and other local breaking news.

It's funny. There are times when I can't stand local news. I feel that some of those broadcast shows truly follow the disparaging adage "If it bleeds, it leads," topping the newscast daily with death and destruction. In their rush to fill airtime and grab viewers from the competition, they are too quick to sensationalize, lazily perpetuating stereotypes and exploiting the very communities that they should be informing.

In their zeal to get a sound bite, they'll head straight to the guy on the block who's up in the middle of the night, putting him on camera even though he's clearly inebriated or struggling with mental health issues.

They don't interview the minister or the teacher. They don't talk to the college student who is home on holiday break or the person working three jobs. They never interview the hardworking guy coming off the night shift who saw the crime go down while everyone else was still in bed, and by not doing so, they do a disservice to the victim and their viewers.

I saw that firsthand, in the clips replayed and pieced together in the newsroom. But I did appreciate the time I spent with Carol. She was very kind and a consummate professional.

Despite all those experiences, in the year after I graduated from college, I wasn't ready to immediately hop onto a professional treadmill. Several of my friends were trying to figure out what they wanted to do, taking in-the-meantime jobs and enjoying freedom from the pressure of exams and résumé building, and I chose to do the same. I had a good time waitressing at TGI Fridays, chatting with customers, delivering orders to the backbeat of the latest pop song by Michael Jackson. My time there even gave me a tiny taste of Hollywood.

Spike Lee was at the vanguard of black cinema, shaking up the industry with his groundbreaking movies on race, love, and the ordinary everyday aspects of black life. He burst onto the scene with *She's Gotta Have It*, but his masterpiece was arguably *Do the Right Thing*, which exploded onto the screen in 1989. It chronicled racial tensions that ignited during one blisteringly hot Brooklyn summer and put a cinematic spotlight on police brutality, which led to the death of a beloved character, Radio Raheem, and in real life continued to snuff out the lives of too many black men and boys.

A year or two later, Spike was back at work in New York shooting *Jungle Fever*, a film about an interracial romance that starred Annabella Sciorra, Wesley Snipes, and Samuel Jackson. Spike was known for giving numerous black actors their first big breaks in the business and also giving folks just milling around in the neighborhoods where he filmed a chance to be bit players in his movies.

Like so many folks working in restaurants, one of the servers at TGI Fridays aspired to be an actor, and she mentioned that she was going to audition to be an extra in Spike Lee's newest film. She invited me to come with her. I thought, "What the heck," and on our day off, we headed uptown.

There were dollies, cranes, and monitors set up all over a block in Harlem. As it turned out, I got chosen and my co-worker didn't, which was a shame because she was looking to be discovered while I wasn't all that interested in being in front of the camera. I wanted to learn the magic that happened behind it. I chatted with the camera operators, asking them a million questions. Then Spike noticed me.

He just picked me out of the crowd. My movie debut entailed my crossing the street with some guy, playing one of several interracial couples ambling down the block. I don't know if I even made the film's final cut, but years later, when I became acquainted with Spike and his wife, Tonya, I reminded him that he'd randomly plucked me out of the background. "I could see that," is all he said. "I could see that." That day on a movie set didn't give me the acting bug, but telling stories, especially about people who were often ignored or misrepresented, or issues that some still considered taboo like interracial relationships, really appealed to me. I'm pretty certain I would have jumped right into journalism to start reporting and narrating those tales. But after my mother's outburst over Christmas, telling me I needed to become a lawyer or a doctor, I wound up taking a dramatic detour and, for a long while, replaced my dream with hers.

I spent more than a decade studying and practicing law, and my journalistic ambitions became a dim flicker. In part that gets back again to the importance of images. I just didn't see anyone quite like me on the screen. There was Oprah Winfrey, the queen of daytime television, but there wasn't any biracial woman steeped in the law as well as journalism anywhere to be found on any channel you turned to. While media still interested me, I wasn't sure I would ever find a place within it.

But I was definitely ready to do something besides practicing law. I wasn't sure where my next opportunity would be, what it might actually look like, but I was open and searching.

Before I could find it, I had two children, my son, Gabriel, and my daughter, Paloma, and put my professional ambitions on hold, choosing to stay home with my kids. After Gabriel was born, I didn't trust anyone with him besides my family. I felt that I wanted him to be old enough to talk before I even entertained the thought of hiring a babysitter, so he could let me know if someone had mistreated him.

When Gabriel was about one, I started dabbling in the law again part-time, working at a securities litigation firm. My aunt Inez came to help us out, and eventually I hired a woman named Junie, who would spend a few hours with Gabriel while I went to the office. But my soul felt unsettled. I had nanny cams everywhere, and in the little time I took to go to work or maybe run an errand, I would check the feed on my phone at least a dozen times.

After I had Paloma, I stayed home until she was about six months old, then I hired an au pair, a young Peruvian woman named Stephanie, who had been a preschool teacher. Gabriel, three and a half years older than his baby sister, was extremely verbal by then, so I was comforted by the fact that he could tell me if something went wrong. I tentatively started venturing out a little more.

The New York chapter of the National Bar Association, the largest organization of African American lawyers in the country, was having a conference. As a young mother thinking of getting back to work, I was interested in networking. And while there, I found someone who was looking for me.

Of course, Stephanie Thompson, a producer for Court TV, didn't know she was specifically looking for me, just as I had no idea who she was or that she would be at the conference. I used to say that it was sheer luck that she and I bumped into each other that rainy day. But someone recently said to me that luck didn't have a thing to do with it. It was preparation meeting opportunity. I believe that now. It wasn't a random, fortuitous twist of fate. I wouldn't have been at that conference if I wasn't eager to network. I knew there would be employers there. I had the flicker of a dream that was getting stronger. I was looking for something.

We started chatting and after a few minutes she said something that would change my life.

"You should be on TV," she said, stopping me in my tracks.

Stephanie told me who she worked for and that the network was seeking lawyers who could analyze and explain the hot legal cases of the day to a cable TV audience. As a journalism major who had become a lawyer, I thought the gig she was seeking to fill sounded perfect. I sized up the situation very quickly and rattled off my education, my credentials, and my interest.

"I have a journalism degree and really want to combine that with my law degree, but I don't quite know how to do it."

Stephanie asked me if I had a head shot and a résumé. "Not with me," I said quickly. But I took her business card and said I would get them to her right away.

I didn't mention that the closest thing I had to a head shot was the senior portrait I'd taken when I graduated from Dominican Academy. That didn't matter. My parents, who had juggled jobs, strategized to get a nicer apartment in a safer neighborhood, and worked overtime to send their daughter to an expensive school, taught me to never be ashamed to seize hold of an opportunity. I wasn't about to play around.

I immediately looked up photographers and booked an appointment with the studio that could get me in the fastest. A day or two later, I headed to the shoot in the pouring rain, had the photographer print my proofs, and when I spotted a shot that I thought was fairly decent, I overnighted it to Stephanie at Court TV with my résumé glued to the back of it.

Within a few weeks, I got a call from one of the network's bookers. She asked if I could come on a couple of days later to appear in a segment. I immediately agreed, then went into a mild panic.

I'd gained maybe sixty pounds when I was pregnant with Paloma, and while I'd lost a little bit of the weight, I was still far heavier than my usual size, and my old clothes didn't fit. I hung up with the booker and called my friend Kathy. When I told her what was happening, she said she was hopping in the car. We were heading to the mall.

Ever the fashionista, and full of knowledge about pretty much anything you can imagine, Kathy remembered reading somewhere that when you were on television, you should wear red because the color popped on-screen. So I bought a red suit and a minimizer bra, to stand out and slim down at the same time. When I showed up at Court TV for that first appearance, which was essentially my tryout, I felt confident.

I was interviewed by two of the network's anchors, Ashleigh Banfield and Jack Ford, but most of my interaction was with Jack. He was incredibly warm and gracious, and our rapport was so strong that though I was supposed to be on for one segment, he asked me if I could stay a few more minutes. Then a few more, and a few more, until I looked up and the show was over. I'd been on air for most of it.

Throughout my academic and professional life, how comfortable I felt in a given place or space had really been my litmus test. If it felt like home, I knew I was in the right place, like I did when I went to SUNY Binghamton and Notre Dame, like I did when I clerked for Judge Bell, researching and writing briefs, and when I was in the courtroom, arguing a case. Offering my opinions and knowledge on that Court TV set, the studio lights shining above me, I felt at ease. I was in the right place.

When the show ended, Jack said to the thousands of viewers who were watching, "It's Asunción's first time here with us. But it won't be her last."

Jack later told me that he had never seen someone who was a total television novice be so perfectly on point and comfortable. He said that he knew that I would be doing TV for the rest of my life.

I felt triumphant after that appearance, elated that I had gotten the opportunity and made the most of it. Before I left the studio, the producers asked what my schedule was for the next month and how many days I could come in and be on set. I didn't really have anything going on professionally, but I played it cooler than that, telling them I could work my calendar around them and was free when they needed me. All I could think as I walked into the cool air was "Wow, this is really happening."

I was over the moon. And then I got a phone call that brought me crashing back to earth.

Stephanie, the au pair, had decided to clip Paloma's nails and in the process snipped off a tiny bit of her pinky. I had been practically levitating off the ground I was so happy about how well the day had gone, but after getting that call, I couldn't have cared less about Court TV. I raced home to check on Paloma. Stephanie was hysterical, and so was I. I was also furious. When I asked her why she'd done it, she said it was because Paloma had scratched her face. She was just trying to keep Paloma from hurting herself.

I called Manny, who told me that the piece of Paloma's fingertip would grow back, which it eventually did. Stephanie was extremely apologetic, Manny calmed me down enough not to fire her, and she continued to work for us for two more years. She turned out to be an incredibly dutiful and caring au pair who became like a member of our family. We're still close. Paloma and Gabriel were the flower children in her wedding.

But that day was bittersweet. It was the beginning of my next juggling act, venturing between two worlds, balancing my career with my personal life and learning to trust someone else to handle things, to watch my children, so that I could fulfill parts of myself that while never as important as my kids, still mattered and helped me to feel whole.

\* \* \*

Although Manny always loved the name Paloma, I wasn't immediately sold on the name. It means "pigeon" in Spanish. There's a park in

Puerto Rico called Parque de las Palomas, where pigeons fly around and literally crap everywhere, and I didn't like the idea of Puerto Ricans looking at my daughter and remarking that she was named for a dirty bird.

Manny didn't get it. "It means dove," he said.

"Ummm, no," I replied. "It means pigeon."

"It's like Paloma Picasso," he insisted, evoking the name of the daughter of the legendary painter.

"Well, yes," I said. "But Puerto Ricans will think we named her Pigeon. I can't have that."

I offered other suggestions. Like Piper. But once again, the Spanish side of the family made that a less than satisfactory choice.

"Peeper?" Nannie Virginia asked, with her accent, when I told her the name I was considering.

"No, *Pie-per*," I said, sounding it out.

"Okay," she said. "Peeper."

We ran it by my mother-in-law, Mara. "Peeper?" said Manny's mother, who was born and raised in a town about two hours north of Madrid, Spain.

That did it. I knocked Piper off the list. I'd take "Pigeon" over "Peeper" any day.

Manny's mom then suggested Pilar, but it just didn't roll off the tongue like Paloma. So, in the end, Paloma was what we went with.

I feel that giving my children Spanish names was very important because it gave them a lifelong connection to their culture. It's why I'm infuriated any time I hear someone call my son Gabe. We named him Gabriel for a reason.

It's harder for folks to abbreviate Paloma's name, though recently I heard someone call her Loma.

"Loma? Really?" I asked her when we got back to the car from her basketball practice. "I know," she said, shaking her head and laughing.

Maybe I get it from my nannie Virginia. Just as I wanted Paloma's and Gabriel's names to reflect their Latino heritage, it was deeply important to her that I be called Asunción.

Mom wanted to name me Bianca, after Bianca Jagger, Mick Jagger's beautiful first wife. Mom loved the Rolling Stones, rushing to buy every one of their new albums, singing "Miss You" as she vacuumed the living room carpet, and standing in line to get tickets to see them when they performed at Yankee Stadium.

But Nannie wasn't down with that. *"Bianca?"*

Nannie wanted me to be named for her beloved sister, who had become a family legend. As Nannie told me, her sister had been very beautiful, and the most handsome and wealthy bachelor on the island of Puerto Rico had fallen in love with her. However, he had left another woman so that he could marry her, and the other woman was devastated and angry. The woman put a spell on my namesake, and the night before she was to be married, as my grandmother's sister tried on her wedding dress and veil, her veil caught on a bedpost, and somehow she fell out of a window and was strangled to death. My grandmother said that even as a child, she saw her sister as a spirit, dressed in her wedding gown, waving goodbye to her, and that she knew the morning of the wedding that her sister was dead.

In addition to honoring her sister's memory, I also think it was very important to Nannie that her kids and grandchildren carry on the culture. I think that was why, though she knew a few words of English, she never spoke them to me. It's why she taught me to make *pasteles* and to season baked chicken with just enough adobo and Sazon adobo seasoning to make the skin crackle and pop. It's why she took me to the botanica and let me watch when she beseeched the ancestors for the many visitors who came to her apartment.

I'm glad that Nannie was happy, but walking through childhood with the name Asunción wasn't the easiest.

Without fail, on the first day of school, or whenever there was a substitute teacher, I'd have to sit there while my name was mangled beyond recognition.

"Where is Susan Parker?" the teacher would say, taking attendance. "Where is Mattie Johnson?"

Then, "Where is Asung, Asoon?" The teacher's brow would wrinkle, as her tongue twisted. Finally, "Miss Cummings. Where is Miss Cummings?"

"Here," I'd say, turning pink with embarrassment.

"How do you say your name? Ascension?"

If it was a substitute teacher I'd never see again, I wouldn't even bother to correct her. "Yeah, Ascensión," I'd reply. "That's fine."

My African American relatives didn't have any problem pronouncing my name, but when I got to college, it tripped up more than a few of my friends. "Do you have a nickname?" some would ask after the third or fourth try. "Well, sometimes people call me Sunshine, or Sunny." With many of those buddies, that stuck.

Now I can pinpoint when I met someone by what they call me. Angelique calls me Asuncion. Kathy and others from my SUNY and sorority days at Notre Dame know me as Sunny, as do Don Lemon, Joy Behar, and the many folks who've known me as a television personality on CNN and ABC.

But one of my own biggest regrets is not starting my television career with my first name, Asunción. That had been my intention, until the day I met Nancy Grace, and I changed my mind.

\* \* \*

Nancy was the anchor I chatted with my third or fourth time appearing on Court TV. She was an earnest, wide-eyed former prosecutor who had a Court TV show called *Closing Arguments* for eleven years, as well as *Nancy Grace* on the HLN network. She had been motivated to pursue the law by personal tragedy—her fiancé, Keith Griffin, had been murdered. She became a television sensation, known for her emotional, sometimes tearful interrogations of folks she felt were on the wrong side of the law and her decibel-bursting debates with her guests.

That first time I was on set with her was like being back in the fifth grade with one of those substitute teachers perplexed by my name. I introduced myself as Asunción Hostin. Nancy kept trying to pronounce it, but it wasn't working out.

During a break, she asked me a question.

"A-soon-cee-yoon," she said, looking like she was in pain. "That's so hard. Do they call you anything else?"

I told her the nickname some of my college buddies used. "You're very good," she said. "Trust me, you should go with Sunny." Like during those school days, I didn't see any reason in that moment to make a big deal about it. "Sure," I replied.

Nancy immediately told one of the staffers to change the chyron, the caption that viewers see at the bottom of the screen, to read as "Sunny" instead of "Asunción," and that was it. I never switched it back. I was now officially, professionally Sunny Hostin.

Nancy and I are still friends, and we've actually laughed about how she essentially christened me. She probably did me a favor.

"Girl, do you think people could remember or pronounce that

name?" she says. And she's right. You can't exactly develop a smooth on-air rapport if the hosts or other commentators are hesitant to address you because they're afraid of stumbling over your name, and producers can't measure your popularity with the audience if people who write or call in aren't sure of the name of the black Latina lawyer they kind of liked.

Still, it doesn't take away the feeling that crops up sometimes that I kind of sold out, that I took a cultural connection and cut it off.

Sunny is not my name. Not that I needed to prove myself, but maybe if I had still gone professionally by Asunción, Ana Navarro wouldn't have doubted who I was, what I was. Maybe *Latina* magazine would have recognized the value of featuring my voice and my story. Maybe it would have been easier for some people to see the multiple parts of me, and my dual identities wouldn't have been so much of a question mark. Maybe.

* * *

I appeared frequently over the next several months on Court TV, rotating between shows, talking about a variety of issues, from the legal loopholes that allowed a disturbed Virginia Tech student, Cho Seung-Hui, to get a weapon and kill thirty-two people at the university in the nation's deadliest school shooting, to the rights of victims to pursue civil charges when their assailants are found not guilty of a crime. I eventually got on the radar of producers at Fox, and one day I got a call asking if I would be interested in appearing on *The O'Reilly Factor*.

Bill O'Reilly was a former CBS correspondent who briefly anchored the tabloid program *Inside Edition* before landing at Fox, then a relatively new network vying to be the conservative counter to what its founders felt were the more liberal voices that dominated television news. O'Reilly was becoming Fox's first major star.

The segment they wanted me for would pit me against another attorney, and the two of us would battle it out over legal issues. Sure, I said, I was game.

To be honest, I had never watched *The O'Reilly Factor*. I don't actually watch a lot of TV news or talk shows, which I know is odd for someone who is a TV personality. When I pick up the remote control, I typically flip to a documentary or look for a movie like *Coming to America* or *Fatal Attraction*, to watch for the tenth or eleventh time. I prefer to get my news by reading it, usually in the *New York Times* or the *Washington Post*. But though I didn't know much about Fox or O'Reilly, I recognized that this was another opportunity, another platform that could help bolster my career.

I arrived at Fox and was introduced to the lawyer I would be debating. Her name was Megyn Kelly. Megyn and I would go back and forth, each of us with our own distinct point of view. Initially, the segment wasn't envisioned as a permanent feature of O'Reilly's show. But it was a great idea that we developed organically, and it proved so popular, it became a mainstay dubbed "Is It Legal?"

Megyn Kelly has had a tremendous fall from grace, as has O'Reilly, who lost his show in the wake of numerous allegations of sexual harassment against him. Megyn, who became a controversial figure when she eventually got her own show at Fox, *The Kelly File*, went on to receive a $69 million contract from NBC, only to separate from the network in 2018 after questioning why people got so upset about blackface. But back in 2006 and 2007, when she and I would debate on *The O'Reilly Factor*, Megyn was just a lawyer like me, getting her first serious stab at television.

One thing I will say about Megyn: She was very smart and diligent. She made me better, keeping me on my toes because I could
not go up against her without being well prepared, and I appreciated
that. On some shows, before an appearance, producers will give you
articles to study and broad themes to stick to, but Megyn and I were
never given talking points. We went at each other with an intellectual
honesty, parrying with well-grounded, evidence-based arguments.

Whenever I would arrive at the studio, I would find Megyn in the
corner, going over legal briefs, and that's the same thing I did. I didn't
want legal CliffsNotes. I wanted the source material. That's what I'd
learned from Judge Bell, who taught me to dig for the rationale given
for passing a law, the circumstances that made it necessary, the debate
on the legislative floor that led to its passage. I wanted to get in the
minds of the people who called for or created the laws and the court
rulings. And Megyn clearly had the same idea. Love her or hate her,
she really did the work.

That's more than I can say for some of the other people I encountered at Fox. The network had not yet become the far-right, full-on
propaganda machine that it is now, a place where I would never appear today, no matter how much exposure I was seeking in order to
build my career. But even a decade ago, I saw things that frankly
shocked me.

When I walked into the building, on the various floors where different shows were taped, I would see a whiteboard with the talking
points of the day. Then, sure enough, those themes would be repeated
throughout every program, all day long. As a legal analyst, those
themes were the basis of every question I was asked.

Some people have their qualms with CNN, feeling it tries too hard
to appease conservatives or veers too far to the left, but I will tell you,

I never saw anything like that whiteboard or script at that network. Never. For all the chatter on the right about "fake news," it is Fox that I found to be rigidly programmed, making its commentators and anchors march in lockstep. That's the same effect it wants to have on its viewers, I believe, almost brainwashing them into what to believe. It was startling.

I appeared on Fox a couple of times a week. I worked with anchor Shepard Smith, one of the few journalists there who I felt gave an objective take on the news. I worked with former *The Five* co-anchor Kimberly Guilfoyle, who eventually left Fox to work for Donald Trump's reelection campaign and became romantically involved with his son Donald Trump Jr., and Gretchen Carlson, who made the first allegations of rampant sexual harassment at the network. I also knew Jeanine Pirro, the controversial judge who has made anti-Muslim comments on air and is a relentless cheerleader for the Trump administration.

I also appeared on Sean Hannity's show numerous times. He was nice, but I didn't find him to be very smart. He was privileged and smug. Instead of going over notes before his show, he would toss a football around with his guests, like it was a romp in the park instead of a workplace. In all my years working with Anderson Cooper, I never saw him throw a football around.

Tucker Carlson didn't have his own show at the time I was working at Fox, but I also thought he was an entitled know-it-all. I remembered thinking, "What makes these people qualified to deliver facts and news to the American public?" With the exception of Shepard Smith, who I believe took his responsibility seriously, I felt many of the anchors and risings stars there epitomized the hypocrisy so prevalent in our society, a double standard that gives opportunities to whites, particularly white men, whether or not they're deserving, while making people of color jump through hoops, dotting every "i" and earning every credential, just to get their foot in the door. It was outrageous.

I met Roger Ailes, the mastermind behind Fox who would eventually lose his perch because of allegations he had abused his power and sexually harassed employees, only once. I didn't get the feeling that he was a fan of mine. My introduction to Bill Shine, Ailes's right-hand man, who was eventually fired from the network as well, and who briefly became the head of communications for Donald Trump's White House, was similarly perfunctory.

Learning the ropes of television, I got myself an agent. But despite my strong work and frequent appearances, I was never made an official Fox contributor. Soon, a new opportunity appeared on the horizon.

Executives at CNN had been watching me. They were interested. Very interested.

Eventually, my agent, Mark Turner, was able to negotiate a very lucrative contract offer with CNN and notified Fox, asking if they wanted to counter.

They did not. They acted like they had been doing me a favor.

"We've given her so much airtime," said Dianne Brandi, one of the network's executives, "and we have so many viewers, Sunny's future is brighter here." As I recall, Dianne then proceeded to offer a quarter of the CNN offer.

Given that this was about business, my agent gave her another chance. "They've offered her that much?" she said, incredulous.

"We love Sunny," Brandi said, "but we are prepared to let her go."

On one of my last days at Fox, I ran into Bill O'Reilly. He'd always been nice to me, though he never seemed able to get my name right. I went by Sunny at that point, which was easy enough. It was Hostin that gave him a hard time. He would say the "o" like he was talking about a Hostess cupcake. I'd correct him all the time, though he never listened.

While Bill had a reputation as a bit of a hothead who screamed a lot, he'd never shown that side to me. I think he knew better. The professional, friendly, but no-nonsense way I carried myself, I believe, put everyone on notice that I wasn't going to tolerate being disrespected. Bill was cordial and would even ask me about my kids from time to time. So I was surprised by the reaction I got when I told him I was heading to CNN.

"Hey, Bill," I said, stopping him in the hall. "I got this offer from CNN. I'm going to be leaving."

"Well, that's your choice, kid," he said. "Do well."

That was it. And he just walked away.

I thought it was flippant and weird. We'd worked together for at least a year. I was told that whenever I appeared on his show, the ratings shot through the roof. I expected he might say, "How much are they offering you? You're so important to the legal segment, you can't go. We've got to figure how to make you stay." But he couldn't have cared less.

I'm not naïve. Kelly was a legal analyst, and I was a legal analyst. She was a lawyer, I was a lawyer, but of course we differed in a significant way. This was Fox News, and Megyn was a blond, white woman, while I was a Boriqua from the Bronx. Race almost indisputably was a factor. But it wasn't the sole reason she paved a future there while I needed to go elsewhere to make my mark.

The truth is that Megyn was ready, and I was still figuring it all out. She was prepared for a greater opportunity, and I wasn't. Not then. Not yet.

When we would be on set, a few feet away from O'Reilly and the green light on the camera would click on, Megyn would command the room. She would take over.

I've now seen this many times, that instance when people realize that they are in a make-or-break moment, and they seize it. They want it. They lean into it.

You have to stand out in the crowd, to burst out beyond it. I think Megyn knew that we were both poised for a breakthrough, and she was determined to be the victor. She would talk faster. She would literally lean into the camera. She would take over the segments. I would speak up, but Megyn constantly outshone me, not because she was smarter, not because she was more polished, but because she wanted it more. I was being a team player. She wanted to be a star.

Megyn and I kept in touch after I left Fox. I sent her an email when she had her first child, and when she appeared years later on *The View* to promote her memoir, I stopped by her dressing room and we hugged and chatted. We had a genial relationship.

Of course she's no longer at NBC. But I respected the hustle that got her there. And I learned from it.

# CHAPTER NINE

# CNN

Unlike Fox, where I ran back and forth from *The O'Reilly Factor* to *Hannity* to programs on the Fox Business network, I had a more formal role at CNN. I was the legal analyst for *American Morning*, the flagship morning show anchored by John Roberts and Kiran Chetry.

I arrived on my first day a bit nervous but also incredibly excited. I had a full-time perch in the morning for the network that had basically revolutionized television, running programming twenty-four hours a day and focusing on news events happening around the world. What I didn't realize as I tried to calm my nerves to get to sleep the night before was that I was about to be thrown to the wolves.

The newsroom was buzzing like a beehive as producers leafed through newspapers and scrolled through online articles, and reporters and camera operators got their marching orders about what assignments they would pursue that day. I wasn't sure where to find my desk amid that swirl of motion, and I soon discovered that no one was the least bit interested in showing me.

I finally asked someone if there was an empty spot where I could set my things, and she absentmindedly pointed me in the direction of a cubicle. I took off my jacket, put down my bag, then continued to stumble around with a million questions dancing around in my head. What time were the morning meetings? Where was I in the rundown of the show? How long was my segment? Was there a script for me to generally follow, or should I just wing it?

I now know when negotiating a contract to include a requirement that a producer be assigned to me so I have some guidance on the lay of the land. But this was my first permanent position in television, and somehow neither my agent nor I had thought about including that provision. Now I had to sink or swim without a lesson or a life jacket.

A producer eventually found me, and I was pulled into a brief meeting about the stories we would be talking about that day. John Roberts, who a few years later would become a White House correspondent for Fox News, resembled a younger version of the legendary CBS anchor Dan Rather. He was serious in his demeanor, but courteous. Kiran was also pleasant, though she was understandably preoccupied, perusing her notes before the show began.

My first segments went well enough. I was well-versed on the topics and had enough experience thanks to Court TV and Fox to not freeze from stage fright when we went live. But the muted, haphazard non-welcome that had greeted me was far from comforting. So far CNN definitely wasn't feeling like home.

CNN marked my first real introduction to the dog-eat-dog nature of television news. Walking the hallways of Fox's headquarters, I'd certainly been aware of how competitive that world could be. But being a half black, half Latino attorney appearing twice a week on a very conservative network didn't exactly make Fox's full-time correspondents and anchors quake in their boots. I doubt they viewed me as a threat, so the knives weren't really pointed my way. At CNN, however, I was under a microscope.

I've always been a pretty quick study, whether I was tackling trigonometry for the first time in the eighth grade or learning the minutiae of gas turbine mergers as an antitrust attorney. But whatever new situation you encounter, whatever new job you take, it helps to have a mentor or an ally who helps you understand the codes and rules governing that particular space. It was one of the great peculiarities of TV news that folks who were willing to pay you a huge salary weren't willing to invest the time to help you do your best and warrant what they were paying you. It was all too clear that CNN was not going to be setting me up for success. Like my daily odyssey as a child to get to and from Dominican Academy, I was going to have to navigate this minefield on my own.

I had a contract, the holy grail for so many, but for months I struggled. No one likes to feel like they don't know what they're doing, but it was even more difficult to fall on my face, more times than I can count, in front of millions of people and to have bosses putting me down, even mocking my performance, while not offering one bit of advice to help me get better.

I'd mastered the back-and-forth banter with other panelists and anchors in my previous television work. But now, dealing with scripted segments and a range of new personalities, I was tripping through the inevitable hiccups that occur when you're doing live TV, like being on air and suddenly having white noise in my ear instead of hearing a report. The folks on set are chattering away, and you can't hear a word. Then a few minutes later, they stop, stare at you, and you have to respond.

Or you write the introduction to your segment, it's fed into the teleprompter, the camera comes on—and the script isn't there. How about you've prepped for the questions the producers have specifically written out so that the appropriate graphics are ready for the piece, and then the anchor pulls a question out of thin air that you're completely unprepared to answer? When I later went to ABC, I would learn that George Stephanopoulos, host of *This Week* and co-anchor of *Good Morning America,* does that all the time.

But perhaps the most humiliating experience of all is actually knowing the answer, then, after the question is posed, your mind suddenly goes blank.

Those are all situations to be expected in live television, especially when you're in the field and participating in a spontaneous, rapid-fire exchange with an anchor back in the studio. But when you're still getting your sea legs, experiences like those shake you. You're like a deer in the headlights, tongue-tied and embarrassed. Almost worse than picturing the millions of strangers watching your humiliation is realizing that your mother and grandmothers are also probably peering into their televisions, their palms sweating and hearts pounding, as they watch you flop around.

I'd always been ambitious, but I didn't have the bare-fanged instincts of many of my new network colleagues. Television is a place where people protect their territory with their lives, ever nervous about losing their position and suspicious of the new face that might steal it away. It soon became clear, as I saw people roll their eyes when I walked by or hesitate to answer questions as basic as where I could find the copier, that there was a lot of jealousy and resentment on the part of people wondering who the hell this new girl making all this money was, for this prominent role, who half the time didn't seem to know what she was doing.

I was trying to tread water in a snake pit. But this was too big an opportunity to abandon without a fight. It was hard to find friends and allies for sure. But I had to try. Slowly, I started building relationships.

I turned first to Kiran, since she and I were working together almost every day. I asked her what I could do better.

"Well, the first thing you need to do is get some mascara," she said without missing a beat.

I wasn't one to wear a lot of makeup, not in the courtroom, not in meetings. My thinking was I wanted people to pay attention to what I had to say, the intellect and skills I brought to the table, not my appearance. Initially, I didn't let the glam squad at CNN do more than put some powder on my nose to take down the shine. But Kiran told me to get over myself and to get my head in the TV game.

"A lot of people watch TV with the volume down," she explained. "You have to look good. This is TV, not radio, honey."

I found the men of color working in the newsroom to be particularly supportive. T. J. Holmes, a handsome young anchor who I'd later work alongside at ABC, had a megawatt smile and a bouncy delivery of the day's headlines that was exuberant and infectious. I naturally have a more subdued persona, but T.J. said that I had to rev viewers up as they started their day, not make them want to go back to sleep. "You've got to have energy in morning television!" he told me.

I also started getting to know Don Lemon. Handsome, with a voice that made you feel like you were wrapped in a cashmere blanket, Don was one of the few people who was welcoming from the first time we said hello. The two of us would chat and grab a cup of coffee from time to time.

Ali Velshi, who eventually left CNN for Al Jazeera and then

MSNBC, was also very friendly, telling me when he thought I'd done a good job and reassuring me when I'd had an on-air flub. "Sometimes a guest throws me something out of left field," he told me once after I'd stumbled through an answer. "I just shift the topic and keep on talking."

I gathered all those bits of advice, filed them in my brain, and began to incorporate them into my work. When I didn't know the answer to a question, I just admitted it and then switched to a subject I did know well. "I'm not sure about that," I'd say, "but let me tell you what I think about this, which is equally important."

If my earpiece went dead, I'd again tell the truth. "I'm sorry, I couldn't hear your question, but let me say . . ."

And I learned to always have notes, known as "blue cards," in front of me in case there was a teleprompter glitch.

I've now been doing TV so long I can quickly recover and even have fun with a memory fail, a tricky question, or a blank teleprompter screen. But learning how to do that thirteen years ago, when all of it was still new, was difficult, even painful. I'm a perfectionist, and each on-air gaffe felt so humiliating.

In addition to having to learn how to recover when there was a hiccup on air, I also soon realized that a single, albeit regular, segment on the morning show wasn't giving me much airtime. With dozens of reporters on staff, there was lots of competition and jostling to be chosen to cover the stories of the day, and I couldn't wait on producers to hand me a juicy segment. I needed to come up with ideas of my own.

Overhearing my colleagues work the phones, I began to recognize that connections were currency, that sources were everything. I likened it to when I was a managing director at a consulting firm. The more business you brought in, the more power you had. Being able to get scoops about an impending indictment or a significant policy change would increase my stature at a network eager to rack up viewers by being first with a riveting story.

I'd hear someone sitting a couple of cubicles away calling the Justice Department and think to myself, "Hold up. I've still got a lot of connections there!" When the Obama administration was deciding who would be the next attorney general, I started phoning my old colleagues to find out what they knew. Working my contacts, I believe I was one of the first people to correctly identify that Eric Holder would become the eighty-second attorney general of the United States. I started to feel I had some power, more than many others around me. And executives and producers at the network took notice.

Next, I began thinking of segments I could pitch that played to my strengths, and I looked for role models, correspondents who got lots of airtime, that I could emulate. One person whose career I thought could be a template for mine was Dr. Sanjay Gupta.

Sanjay was a neurosurgeon who had carved out a role as CNN's medical expert. He's been a fixture on a number of programs and has hosted specials and shows where he's offered his expertise on everything from the Ebola virus to the questionable nature of medical assessments given by the physicians tending to Donald Trump.

Though Jeffrey Toobin, a bespectacled Harvard-trained attorney and writer for the *New Yorker,* had the legal analyst slot pretty much locked down during CNN's evening hours, I thought I could model a legal role similar to Gupta's medical one for the network's daytime programs.

One day I pulled Sanjay to the side. I told him that my in-box was filled with emails from viewers asking my advice on issues they were dealing with, whether it was a dispute with a neighbor or a testy relationship with an office manager. Sanjay had a segment called "Paging Dr. Gupta," where he addressed viewers' questions about health. How about I pitch a similar segment called "Sunny's Law"?

Sanjay was amazing, supportive, and generous. He thought it was a great idea. And, to my delight and I'll admit my surprise, some of our producers agreed. Soon I had a branded segment of my own.

"Sunny's Law" resonated with viewers. Soon, I asked to have my own producer, who could help me choose the questions and better shape the segments. I was assigned Saundra Booker, a black woman who was a news veteran. She had a great feel for the types of topics that would resonate with the largest slice of our audience.

I also decided that I wanted to do more than just offer analysis or answer viewer questions. I wanted to be a correspondent, to go out in the field and file reports. Again, I'd watched Sanjay do that, and I knew that I was as familiar with courtrooms as Sanjay was with an operating room.

Since the medicine and the law often intersected, like in the realm of medical malpractice lawsuits, Sanjay was game for us to report some pieces together. One segment I did that touched on both fields was a report on the plight of Filipino nurses who had filed a lawsuit against the nursing home where they worked.

I was going to have to track—TV speak for recording a script, something I hadn't done much of before, since I was usually either offering live commentary or working with another reporter, like Sanjay, who would narrate the story. Now that I would be the one recording the track, I figured I would use an inflection similar to the authoritative tone I'd fall into while making an argument in court or conducting an interview. But tracking was a completely different skill, requiring me to know how to emphasize certain words in a way that I wasn't used to.

I'd do a take, and the producers would toss it back. They didn't try to be kind. "This is trash. Do it again."

I would, then I'd hear more criticism. "You're really not good at this," one producer told me.

"I know," I responded dejectedly. "Do you have any tips?"

"I don't have the time," he'd say. "But if you can't get this done, we may have to nix the piece."

I was worried, exasperated, and really needing help. It suddenly flashed in my mind that maybe Soledad O'Brien could give me some advice.

Like me, Soledad was a black Latina, but unlike me, she seemed to have found her footing at CNN, carving out a niche as a correspondent as well as an anchor. We weren't necessarily friends, but she was always nice, greeting me when we passed each other in the studio or met up in the ladies' room.

I went and knocked on her door. "I need your help," I said.

Soledad didn't hesitate. She walked with me to an editing booth, and together we got to work.

"Think of it like this," she explained. "You've got only fifty dollars, which is equal to a paragraph. Every word that's important enough to spend ten dollars on, that's the word you circle, that's the word you punch. That's the word you punctuate as you speak."

Making her tutorial so simple to understand really gave me a road map I felt I could follow. I did the track exactly the way she said. The producers didn't praise the fact that I'd finally gotten it right, but they didn't tell me to do it again. The piece finally aired, and it and I got a lot of positive feedback from viewers.

I don't necessarily think the indifference I encountered at CNN was motivated by conscious racism. Television news, with its promise of influence, celebrity, and a lot of money, is ferociously competitive, and some are savvier about cultivating management and getting themselves in the spotlight than others. But black and brown folks remain a minority in newsrooms, and because we don't look like the majority of people who've typically been the medium's anchors and stars, we're not always viewed as having the same potential or deserving of the same mentoring.

During my career, I've seen many people get weeks, sometimes months, of media training and coaching, even though many of those reporters and hosts were coming from stations in major cities, or even another network, and presumably should have already known how to track a piece or write a tight script. I've seen reporters show up unprepared day after day or be a terrible fit for a position, yet stumble higher and higher up the ladder as they were given opportunity after opportunity. Or perhaps they would be handed a lofty position with no prior experience and have their hand held until they slowly began to improve.

I wasn't afforded that teaching, that understanding, that forgiveness—ever. And the truth was, it was infuriating.

* * *

Television is a revolving door, as shows are reshaped or replaced to meet the ever-changing moods of a fickle audience. Morning television, the real moneymaker in broadcast and cable news, can be particularly volatile. One day, after months of tepid ratings, the team at *American Morning* was told that our show would be getting a new executive producer.

Janelle Rodriguez was a Puerto Rican woman with a lot of experience under her belt, and a breath of fresh air. Her energy and ideas about the show's topics and pacing made it feel more dynamic and inviting. I liked her, I liked the changes she implemented, and I felt I was finally hitting my stride, scoring scoops, reporting assignments, and, of course, still steadily providing legal analysis.

At the end of Janelle's first season, we were called in to meet with her. I looked forward to having a sit-down. Though I was more comfortable, I knew that I was still a novice with a lot more to learn. I wanted to take on a potentially bigger role with the show, maybe even occasionally fill in when John or Kiran were off. So the morning Janelle and I were set to chat, I grabbed a legal pad and jotted down some of the points I wanted to discuss.

I didn't want to give a sob story, but since Janelle hadn't been around when I first got hired, I did want to tell her that I'd struggled and hadn't gotten much feedback. I was interested in media training and any other tips she had to give.

Janelle gave me a puzzled look as I sat down. "Oh," she said, looking bemused. "You're ready. You brought your legal pad and everything."

"Yes," I said, ever the eager A student. "I'd like to bounce some ideas off you."

It quickly became clear that she wasn't interested in my ideas, or anything else I had to say.

"I brought you in here to tell you that we've decided not to renew your deal," she said. She seemed to be on the verge of laughing at how far left this meeting had strayed from what I had been expecting.

"Pardon me," I asked, the blood draining from my face and rushing down to my feet.

"I had to make some cuts," she replied matter-of-factly. "And you're one of them."

There I was, bright-eyed and bushy-tailed, thinking this meeting was like the performance evaluations I'd had throughout my legal career, when my boss and I discussed goals and ways I could improve. Instead, I was being fired.

That's TV, chaotic and cold. One day you're on top of the world, clutching a big contract, and a year later, you're out of a job. If a show isn't doing well, there's too much money on the line to give it years to "find" an audience. Today I understand that's the way it works, and years later, once Janelle had moved on to a position at NBC, she actually reached out to me about possible opportunities there. But that day in 2009, I felt foolish. It was really the first time professionally that I had failed at something. I packed up my yellow legal pad, the photos of Manny and my kids that covered my desk, and went home.

I was devastated, partly because I really hadn't seen it coming. After that flub-filled first year, I'd started to get smoother on air and better at handling mishaps when they happened. The teaching I'd gotten from folks like T.J., Kiran, and Soledad had definitely helped, and it had been a while since I'd gotten negative feedback from producers or others I worked with.

But my television career appeared to be over. Maybe Mom had been right after all.

*  *  *

I went back to doing some legal work part-time. Then, about a year after I was fired from CNN, I got a call. Court TV had been rebranded truTV and had a new management team led by a man named Ken Jautz. One of its programming changes was to create a daily feature dubbed *In Session,* in which legal cases and breaking news were discussed for a few hours. The producer who rang me up said she remembered that I'd been one of the stars in Court TV's early days. Would I be interested in coming back as one of the network's legal analysts?

Court TV was where I'd started, the place that gave me my first shot at a years-long dream. I'd really enjoyed it. And I even missed the frenetic hustle of CNN, as much as a trial by fire my time there had been. I very much wanted to get back into TV. I didn't have to think twice. I said yes.

TruTV was owned by Time Warner, CNN's parent company, so both networks were housed in the same building. It hurt a little going back there, given how I'd left CNN and knowing that I now would be working for an entity that was a less prestigious part of the network family. But getting off at a different floor was only one reminder of how far I had fallen.

Television, with its long hours and savage politics, is in many ways not as glamorous as people think. But my latest gig didn't even have the glossy veneer. I wasn't making anywhere near the money I had before. And instead of being picked up every morning by a town car and whisked to a spacious, beautiful set, I was driving myself to the Time Warner Center, CNN's massive headquarters in Columbus Circle.

I would give my commentary in what was called a flash studio. Sometimes I'd sit by myself, but often I'd be squeezed behind a tiny desk with another attorney. The camera would move back and forth between the two of us to make it seem like we were offering our analysis from two different rooms when we were actually almost sitting on each other's lap.

It was a humbling experience to be sure. But I took the job because

I wanted to keep that on-air muscle working. I felt that however spartan or bleak the surroundings, I was fortunate to be able to still have my voice heard on the legal cases of the day. Occasionally I would be flown down to CNN's Atlanta headquarters, and I also appeared on HLN, a sister network to Tru and CNN. At HLN, I was reunited with my old friend Nancy Grace.

The story of the solitary figure who pulls him- or herself up by the bootstraps to ultimately reach the heights of success is a defining American myth. But as central to American identity as that story may be, it is indeed a fairy tale. I know firsthand that as prepared and hardworking as you might be, it's critical, maybe even the difference maker, that you have people in your life who are willing to give you a shot, people who will give you the stage to show what you're capable of, whether it was Ms. Lopez recommending I be skipped to fifth grade, or the head of Notre Dame's alumni association recommending me for a clerkship in Judge Bell's chambers. Nancy too was always one of my cheerleaders.

She gave me my first chance to anchor, telling the producers to let me fill in for her when she went on vacation. I'll always be grateful to her for believing in me and allowing me to stretch and grow. Nancy's style wasn't everyone's cup of tea, but she had a big following, and when I substituted for her, I was able to maintain the ratings, which was typically all that mattered to programming executives. It proved I could potentially carry a show.

I began to think more about becoming an anchor. But whenever the network sought out new hosts, instead of giving me a tryout, they'd have me sit in with the typically male job candidates in my standard role as an analyst. I was basically helping the other person with their audition. I found that unbelievably insulting. I'd filled in several times for Nancy Grace, arguably HLN's biggest star, and hadn't lost viewers. I wanted a tryout as well.

I called one of the top executives, Tim Mallon, to ask him what was going on. "You're good at what you're doing," he said, basically telling me I needed to stick with the tried and true.

"I understand that," I said. "But I still want to audition."

I kept asking, and maybe, to get me to quiet down, I finally got one. The network actually flew me from New York to Atlanta for it, and I did a great job, if I do say so myself. But apparently, I was the only one who believed that. Tim told me that I didn't have what it took to be the full-time host of a show. In short, he was telling me I needed to stay in my lane.

I still have that audition tape. Since becoming a co-host of *The View*, I've actually rewatched it. And you know what? It was good. Really good. The executive with the power to make things happen simply didn't agree. He couldn't see it, and that's okay. It's a subjective business. But I knew that I couldn't go back to sitting at a desk in that cramped studio, a camera toggling back and forth. That moment of rejection lit a fire under me. I had to find a new way.

By then, I'd been at Tru for a couple of years. I'd been a team player, doing good work. Now I wanted more. I went to Ken Jautz, to his counterparts at the mother network, and also to my agent and told them all that I wanted to move on from *In Session* and become a full-time analyst for CNN and HLN instead.

I also started seeking out folks like Anderson Cooper and Don Lemon, who both had their own shows. I'd appeared on their programs to offer legal commentary. Now I asked them if they would have me on more often.

My stint at Tru also didn't prevent me from appearing elsewhere, so I offered legal analysis from time to time on the weekend edition of the *Today* show, as well as programs on CBS. I had conversations with executives at MSNBC to feel out potential opportunities there.

I also had people like Nancy Grace and Soledad who had my back, vouching for my talent, telling producers, "Hey. Sunny can do this."

Luck is preparation that has a rendezvous with opportunity. I think all of it, my popping up on other networks, my increasing appearances on CNN shows, the co-signing I got from well-respected colleagues, helped to usher me back fully into the fold. My agent got the call. I would officially again be part of CNN.

* * *

Out in the wider world, as my TV career progressed, I started to get recognized more and more. I'd be walking down the street and see a passerby suddenly do a double take. Or, occasionally, I'd go out to get a sandwich only to be shyly approached by another customer asking if I was "that lawyer on TV."

But in our society, where entrepreneurs can have the police called on them at a Starbucks because the barista is wary of two black men sitting and not ordering anything, in a country where black Ivy League college students are profiled and pinned down by security guards, in a nation where the police are called to shut down black people grilling in an Oakland park, being a television personality didn't protect me from the bigotry that continues to dog black and brown people, no matter what we've achieved.

Like the not-so-long-ago day when I encountered a white salesclerk at an upscale department store in Westchester. From the moment I walked in, it was so clear that she was following me.

It was a fall afternoon. I had a much-needed day off from CNN, and my mother and I drove from our home to a store in Westchester to do a little shopping.

As soon as we stepped off the elevator, the saleswoman appeared. She was young, white, and intent on shadowing me.

When I moved, she moved. Where I went, she followed. If I'd walked out of that store with overflowing shopping bags, her paycheck would have likely gotten a significant boost since most salespeople depend on commissions. But she couldn't have cared less about giving me any assistance.

I fingered dresses while she hovered a few feet away, shuffling shirts and slacks while keeping tabs on me out of the corner of her eye. I pulled a pair of slim-fitting denim pants out of a stack, wanting to try them on. But neither she nor any other employee approached me.

I wandered over to the section of the floor dedicated to the designer label Max Mara, grabbed a slouchy pair of pants, and still no one on the sales floor asked what else I might be interested in or offered to get me a dressing room key. I stood fidgeting as I watched other shoppers—white women—stream in and out of the fitting area. Meanwhile, my shadow continued to linger nearby. I nudged my mother.

"This woman has been watching me ever since we came up to this floor," I said. Mom quickly glanced behind me. "Oh," Mom said, waving her hand as if to brush away a pesky gnat—or a thought she didn't want to entertain. "She probably just recognizes you from TV."

But I knew that sales associate didn't want an autograph. Like so many black women, I'd grown used to the daily slights and indignities, the myriad microaggressions, heaped on us no matter our accomplishments.

I usually made a point of dressing up a bit when I went into the luxury department stores and boutiques along Fifth and Madison Avenues, slipping on a pair of Manolo Blahnik pumps, wrapping my shoulders in a Chanel scarf, just so I'd be given the benefit of the doubt that I did have money to spend, that I did deserve proper service, that I did deserve respect.

But when I decided to go shopping that morning, I just wanted to be comfortable. I'd tossed on a baseball cap, a pair of Lululemon athletic pants, and an oversize shirt, then headed out the door.

That saleswoman didn't see Sunny Hostin, a former federal prosecutor with a permanent spot on the biggest cable news channel in the world. She saw a casually dressed black women whom she probably figured had no money and certainly had no business strolling the perfumed aisles of that store. I was at best a customer who wouldn't be worth the time to wait on, and at worst a potential shoplifter for whom she'd have to summon security or even the police.

I'd had enough. "Excuse me," I said to the saleswoman, who by then had tailed me for nearly half an hour. My voice was calm but firm. "I don't need any help. And I also don't need you following me."

She looked startled, her eyes widening. Then, without offering an excuse or an apology, she turned and slunk away.

Just as my being in the public eye didn't shield me from prejudice and profiling, my dual identities continued to be questioned even though I was very open about the fact that I was black and Puerto Rican when I spoke on TV. CNN Español, for instance, barely gave me the time of day.

I'm not sure where the disconnect came from. My office was practically next door to the collection of rooms that was CNN Español's New York headquarters. Sometimes I told myself that maybe it all would have been clearer if I hadn't made the name change, if Asunción had been on the plate on my door instead of the anglicized, simplified "Sunny."

But I was a member of the National Association of Hispanic Journalists, attending its national conferences and the occasional local chapter meeting, where I was sure some of CNN Español's team members had seen me. And I knew that black people at CNN were definitely aware of my being part Latino. The painful reality is that CNN Español's initial failure to acknowledge me probably had more to do with the same narrow assumptions about what a Latina is supposed to look like that I'd experienced throughout my life. The producers over there were much quicker to claim my office mate, Jean Casarez, who was also half Latino but very fair-skinned and blue-eyed, and Soledad, whose mother was Afro Cuban.

Again, it was like I had to offer some kind of proof.

"Are you Latina?" a producer finally asked me one afternoon, after she overheard me chatting in Spanish with Jean. It was like she'd suddenly heard a puppy recite Shakespeare.

When I told her that my mother was from Puerto Rico, she and others at CNN Español finally became more inviting, saying they would love for me to do some reports for them.

Years later, when I went to ABC, I would have a dramatically different experience with my Latino colleagues. By then, the broader community had embraced me, inviting me to be an ambassador to the Puerto Rican Day Parade no less. I would get shout-outs walking down the street, with strangers greeting me in Spanish. And inside the network's halls, it was literally *"Hola, amiga!"* I was interviewed in Spanish on WABC, ABC's local New York affiliate, and I became pretty tight with Latina co-workers like Cecilia Vega. But at CNN, that kind of recognition took a while to take hold.

Still, given the crazy contradictions of color, race, and perception, my lighter skin and hair did at times give me a buffer from some of the slings that my darker-complexioned peers had to endure.

I don't have specific anecdotes to share about how I was chosen for a slot or given consideration for an opportunity, because whites felt more comfortable looking at me than someone who had ebony skin or hair twisted into locks. But I don't fool myself. The favoritism doesn't have to be that explicit for me to know I benefit from it.

There are white people who chafe when white privilege is mentioned, saying they've never been handed anything because of their race. But to me, that's not what white privilege is. Rather, the benefits arise from absence, the absence of bigotry and unfair barriers; the absence of stereotypes and diminished expectations. Very simply, privilege is the freedom that comes from not facing discrimination because you're white.

And just as there is a thing known as white privilege, there is privilege afforded to those who are lighter-skinned too. I've certainly never had anyone white ask to touch my hair like they've asked Kathy and some of my other browner-skinned friends, as if their textured tendrils were some kind of space oddity. Not having to contemplate how to respond to something that is not malevolent is certainly othering, as well as far more negative assumptions or perceptions is definitely a privilege. It would be ridiculous for me to think such advantages don't exist in a society as color conscious as ours, even if I sometimes also face anti-black bigotry. I know that it does.

* * *

While I appreciated being back on CNN as an analyst, I'd grown a lot in the five years since I'd first walked into the studio not even knowing where I should sit. I was more knowledgeable, more confident, and more sure of where I wanted my career to go. I wanted to have more power to decide and shape stories, to get pieces about inequitable sentencing and racial profiling on the air. To have that kind of say-so, I needed to be an anchor.

I began to aggressively lobby for the chance to fill in when any of the network's anchors were off. But some of my female peers were not as generous as Nancy Grace. Not by a mile. One anchor, who continues to have her own show and who at the time I thought was a friend, specifically told the producers that she didn't want me to fill in when she was away.

Another time, I finally got on the schedule to substitute for a female host who was going to be on vacation for a week. When she found out I would be her replacement, she actually canceled her time off with her children so that I couldn't sit in her anchor's chair for a few days.

On the one hand, it bothered me that these women would behave so selfishly, particularly when I'd really trusted one of them. On the other hand, I thought that I must be pretty good if they were that threatened. Maybe I was on to something.

Just as I've cherished the many mentors and friends who've supported me throughout my life, I've also remembered people like those two colleagues who tried to undermine me, and I've resolved to never be like them. I believe the universe has enough bounty for all of us to share, so no one needs to hoard our good fortune. And I also believe that if I'm meant to have something, to occupy a certain space, no one else can replace me. So, because it's the right thing to do, and it's the model I've seen from family members, teachers, and mentors all my life, I try to pass the baton and help others in any way I can.

That is unless you have shown me that you are a different kind of person. That so-called friend of mine who asked that I not be given the chance to fill in for her called me after I'd become a co-host of *The View*. She asked if I would recommend her as a possible guest host when one of the other hosts was off. I told her I would see, though I don't plan to. But if Nancy Grace, Kiran Chetry, or one of the many other women I've known made the same request, I'd suggest them in a minute.

Between the producer who gave me a tryout but was unimpressed, and the anchors who blocked me from filling in on their shows, becoming even a substitute anchor was proving elusive. I began to have doubts, fretting about what I was doing wrong, about how I could break through. I looked again for role models. There weren't many. But Soledad was the most like me.

Soledad was one of the few people who had been kind enough to help me when I first started at CNN, and she remained someone I knew I could call on. She was the only woman on the air who had a somewhat similar ethnic background. And Soledad actually had been an anchor, co-hosting *American Morning* and then the early day program *Starting Point*. She also was the anchor and force behind *Black in America*, a documentary series that chronicled various aspects of life in the black community

CNN didn't seem particularly enamored of me, this Afro Latina from the Bronx who felt she had to use her platform to be a truth teller and could get a little heated when etching an argument. Soledad played it safer on air, with her "just the facts," non-opinionated delivery. Maybe Soledad was the kind of woman of color, the type of persona, the powers that be preferred. Maybe, I figured, I should try to be more like her.

My hair is naturally a lighter shade of brown, but I dyed it to resemble Soledad's dark, shoulder-length tresses. Even though my voice is much deeper, I began to mimic Soledad's clipped tones. And I held back from giving my opinion, engaging in the both-sidesism that unfortunately has become the hallmark of too many cable news shows in the age of Donald Trump, as mainstream journalists strain to be deemed "fair" by conservatives who don't follow the same playbook.

I don't know what I was thinking. It was a bit ridiculous, because I'm so different from Soledad. And it's funny, because since we both left CNN, Soledad has become quite the firebrand on Twitter. I tell her jokingly that she's me now, but worse. At the time, however, I thought that the best route to achieving my goals was becoming a clone of Soledad, circa 2010.

What I know now is that what works in television is authenticity. The talents who stand out, who become stars, are uniquely who they are, occupying a space that only they can fill.

Think about it. How many androgynous, blond lesbians like Ellen DeGeneres have you seen on TV? Oprah Winfrey stood out, not just because she was a great interviewer but because she was a *Super Soul Sunday* font of inspiration who didn't look like anyone else on the airwaves during the daytime. The same can be said about Arsenio Hall, Phil Donahue, Anderson Cooper, and even Wendy Williams, all of whom had a unique look, voice, or style of interviewing that made them a refreshing change from the standard fare.

What I didn't realize years ago was that my roots in the South Bronx, watered by bloodlines that flowed through Georgia and Puerto Rico, made me unique. I'm a bookworm who keeps it real, a lawyer whose refinement is spiced with a bit of swagger, a bicultural woman who's lived a life in the gray. I look different and speak differently from just about anyone else who looms on the screen when you click your remote. And that's a package worth watching.

I didn't understand that when I went to the hairdresser to get my hair dyed, when I modulated my voice and muffled my opinions. But soon I would stop trying to be Soledad and become solidly, defiantly Sunny.

There was a young man who made me dig deep. He compelled me to shout about the injustice that led his killer to literally get away with murder. He made me summon the South Bronx, the prosecutor, and the mother all tangled up inside me to proclaim the humanity of a black child taken too soon. I had to find my voice to call up his memory as often as I was able. His name was Trayvon.

# CHAPTER TEN

# TRAYVON

was in my office one day in March 2012 when my cell phone rang. It was Benjamin Crump.

I had been friends with Ben, a driven, passionate civil rights lawyer based in Tallahassee, Florida, for several years. We'd been on panels together at conferences convened by the National Bar Association and he'd occasionally call and cheer me on after he'd seen me dissect a case in my role as a legal analyst at CNN.

Ben's warm manner and distinct Southern drawl might lead some to underestimate him, but he is a brilliant lawyer and strategist. He told me that he was representing a couple, Sybrina Fulton and Tracy Martin, who were trying to draw attention to a tragedy that is every black parent's nightmare.

On February 26, 2012, their son, Trayvon, was walking back from a store in Sanford, Florida, when he encountered George Zimmerman, a neighborhood watchman and wannabe cop. Zimmerman, who'd called the local police because he viewed Trayvon, a seventeen-year-old boy carrying candy and iced tea, as "suspicious," shot and killed Trayvon, claiming he'd done so in an act of self-defense. The cops didn't even bother to arrest Zimmerman based on a peculiar law dubbed "Stand Your Ground," in which a person could justify killing someone by saying they feared for their lives, though apparently not enough to try to run away first.

These were the types of injustices Ben had been fighting against his entire legal career, case by case, outrage by outrage. In Sanford, a storm of protest fueled by pain and grief was gathering. But Ben knew this was bigger than the Miami suburbs, bigger than Florida, bigger than the south. It was emblematic of a national epidemic, and it deserved national attention.

To help garner that kind of coverage, Ben enlisted a white public relations guy, Ryan Julison, who immediately asked him if he knew any reporters working at a national outlet. Ben mentioned that he knew Roland Martin, a CNN contributor who eventually got his own show on the African American–owned network TV One, and he knew me.

I didn't need convincing. I was viscerally connected to the story because I had a nine-year-old black son of my own. I kept thinking, "What if that were Gabriel lying on the ground, Skittles and iced tea spilled beside him, shot and left to die alone for no other reason than that he had black skin?"

But you didn't have to be a black mother to be swept up in the story of Trayvon. What I think Ben realized early on is that Trayvon's mother, Sybrina, so elegant and dignified in the midst of unbearable grief, was the woman who could propel this too-common tragedy into the national consciousness. He knew that any other mother, whatever her ethnicity, could look into Sybrina's mournful eyes and immediately identify with her pain.

Ben brought Sybrina to CNN not long after he'd phoned me, and I'll never forget our first encounter. Sybrina's grace called to mind Mamie Till, the mother of Emmett, who in 1955 allowed *Jet* magazine to publish pictures of her beloved only child's mutilated body so the world could see the savagery of bigotry firsthand. She made me think of Coretta Scott King, beatific at the funeral of her husband, Martin Luther King Jr., and Jackie Kennedy, numb and anguished in her bloodstained Chanel suit.

In all the months and years to come, through the effort to finally get Zimmerman arrested, the trial that followed, and its soul-crushing aftermath, I don't believe I ever saw Sybrina angry, though she certainly had the right to be. She displayed the kind of dignity we all would hope to have if we had to endure such a heartbreaking ordeal, though most of us doubt that we could.

Sybrina just wanted justice for her baby. She just wanted to tell his story, for the world to know how precious he was. She showed me photographs of Trayvon riding a horse and beaming as he prepared to ski down an icy slope. The truth was that black children, as innocent and deserving of tenderness and care as any others, were often ignored or forgotten, whether they went missing in broad daylight or lost their lives to an assailant's bullet. Sybrina was determined that would not be the case for her son.

I immediately began approaching producers about this story, insisting that we had to cover it. And they immediately dismissed me. It was a local story, they said. They weren't interested.

I'd grown used to pitching a lot of pieces that were shut down because they were viewed as too inside baseball, burrowing into the nitty-gritty details of the law. This wasn't that, so I kept pressing. This was a story about race and guns and the law, issues that infected and affected the entire nation. A white man had killed a black child, something that happened too often in this country. The Stand Your Ground law that gave people license to take another's life virtually without consequence was on the books in several states.

"Think about it," I said. "A black boy gets shot by a fake cop. He doesn't have a gun or a weapon of any kind. And in Florida, you can just say you're afraid of somebody, shoot them, and not even be arrested. This is crazy."

Most of the producers I spoke to said they still didn't see it as a national piece. But thankfully Anderson Cooper and his producer, Charlie Moore, did.

You didn't come much closer to American royalty than Anderson Cooper, son of the socialite and designer Gloria Vanderbilt. Yet he was exceedingly down-to-earth, kind, and always supportive of my career.

Anderson had done hard work out in the field, covering the human and psychological toll of Hurricane Katrina and other gut-wrenching stories with a gentle humanity, really working for the success he attained despite being born into wealth. It also made him confident, knowing he'd paid his dues to earn his place in the TV news pantheon when someone else with his background might not have felt the need to. Rather than walling himself off, playing the entitled superstar, he was always willing to help someone else.

Charlie, Anderson's producer, was also incredibly talented and a font of support. In life there are those people who will give you a little space to try things, to make mistakes, and to stretch your muscles. Charlie and Anderson were two of those people for me. So when I told them Trayvon's story, they said if I needed a place to air it, I could do it on Anderson's prime-time show *Anderson Cooper 360°*.

I proceeded to put together the package, using the photos and anecdotes that Sybrina and Tracy had shared with me. Trayvon was very good at science and math, and he wanted to be either an aviation technician or pilot. He'd actually spent four years attending a program called Build and Soar run by Capt. Barrington Irving, the first black pilot to fly around the world. After recording interviews with Trayvon's family and assembling the pictures and other elements, the piece was ready to be stitched together by a producer.

I remember walking into the room as she was putting on the final touches. We'd had photographs showing Trayvon as the child he was, vacationing with his family, smiling as he spent time with friends. But when I looked, most of those pictures weren't in the piece. Instead, she had decided to use, again and again, a picture that was being circulated by some local media outlets.

The picture showed Trayvon wearing the type of metal grill on his teeth that had briefly been an unfortunate fashion statement among teenagers, both black and white. That grill, that photo, transformed Trayvon into someone that he wasn't. It soothed viewers who just didn't want to believe or acknowledge that they lived in a country where in the twenty-first century, an innocent black teenager could be murdered for doing nothing more than walking through a suburban community, wearing a hoodie, and clutching a bag of candy. And his killer could walk away scot-free.

The producer of the piece was someone I'd worked with before, and I don't think her actions were driven by malice. For all I knew, that was the third piece she'd edited that day, she was heading into overtime, and she simply went for the picture she'd seen everywhere else, not giving a thought to the narrative that it helped convey. Even more likely, this was just another example of black boys having been so debased by local news programmers that even well-meaning people weren't sensitive to such inaccurate, destructive images. But ignorance was no excuse. I wasn't having it.

"This picture misrepresents who Trayvon was," I said. "I have pictures that no one else has seen. We need to use those."

She gave me a puzzled look but didn't fight me. She recut the piece, added my pictures, and left the photo taken of Trayvon on perhaps the one day that he'd ever donned a grill on the cutting-room floor.

That moment demonstrated to me a truth I had absorbed as a prosecutor, interviewing the victims of sex crimes, or when I threw out a confession obtained from a man who'd clearly been a victim of police brutality. You had to be in the room.

Because I was in the room, the piece that aired portrayed the real Trayvon, the full Trayvon, the innocent child who loved mathematics and planes, instead of reducing him to a single snapshot that would have led many to wrongly see him as a thug. You had to be in the room to offer perspective, to make the final decision. All of us being there got one of the first national pieces detailing the tragedy that had occurred that day in Sanford, Florida, on the air. And that helped to light a fire.

Trayvon's mother later thanked me for the piece. But the battle it had taken to do that report was a harbinger of what was to come. I would have to keep fighting to tell Trayvon's story, to shape his narrative. And for that, I would have to summon my voice.

* * *

Initially I was still trying to be Soledad. I tried to speak about the Trayvon Martin story mostly within its legal parameters, the fact that an arrest finally happened, the reasoning given for Stand Your Ground laws, and what the defense was arguing versus the prosecution.

But my worldview was shifting. I had been outraged when I read stories about Amadou Diallo, a Guinean immigrant who was shot forty-one times by four police officers while standing in the vestibule outside his Bronx apartment in February 1999. I'd also been deeply saddened when I heard about Sean Bell, another black man murdered, this time by undercover and plainclothes policemen on the eve of his wedding in November 2006.

Trayvon, however, was different. It was different because a friend had brought his story into my life. It was different because rather than reading about it in the newspaper, I was reporting on it myself for cable news. And most of all, it was different because I was a mother and he'd been a child. I could envision myself in Sybrina's shoes, though it was a thought I didn't dare think about for too long.

I became keenly aware of something I'd long suspected but now was observing in sharp relief. I was seeing how the media could twist a story, reframing it to meet the prejudices and misperceptions of the people in charge or of the people whose viewership and readership the outlet coveted most.

Trayvon's cherubic face was everywhere, marred by that metal grill. There were stories about him perhaps having once tried marijuana, like so many teenagers, white and black. But the allegation was clearly being flung at Trayvon to paint him as somehow dangerous and deserving of suspicion and fear.

Yet you weren't hearing about Zimmerman's background, how he'd wanted but never been able to get into the police academy, how he had a history of domestic violence and a spotty employment history. It was like reporters had forgotten who was the suspect and who was the victim. Rather than focusing on the man who'd killed a teenager, major news outlets were assassinating the character of a child who was just trying to get home when he was attacked and lost his life.

Zimmerman's trial began in 2013, and I was assigned to cover it, reporting from the courtroom as well as offering analysis. I packed my suitcase, headed to Florida with a producer and technical crew, and basically moved into a Marriott hotel.

In many ways the trial, and the ugliness and racism it revealed, was a watershed moment for the country. It certainly was for me. If my worldview had shifted in those first weeks when I was reporting on the story from CNN's New York studio, then sitting in that courtroom, my worldview began to crack.

I'd been living the American dream. I had risen from the poverty and chaos of a childhood in the projects of the South Bronx, to graduate from an exclusive girls' academy and ultimately law school. I believed that hard work, along with the support of a wonderful family and a solid education, could level the playing field and make a person successful because my life had given me no reason to doubt it. I was an attorney married to an orthopedic surgeon who was the most incredible father to our two beautiful children. I was working in television, my lifelong dream. And I'd experienced history, seeing the election of Barack Obama, the nation's first black president.

All of that led me to believe that I lived in a country, in a world, where more often than not, people really were judged by the content of their character rather than the color of their skin. It may seem unbelievably naïve that I thought that way. And I certainly had experienced my own brushes with bigotry, but I really did think that hard work, and goodwill, ultimately paid off.

Now I was sitting in a courtroom, a mother of two children of color, listening to the story of how this kid was racially profiled by a vigilante and killed for doing nothing other than just being who he was. It was shattering, and the racist ideas, perceptions, and laws that led to his death were undeniable.

I started realizing that I was the right person to tell this story, that my platform gave me power and that I had an obligation to use it to speak for Trayvon, who couldn't speak for himself, and for Sybrina and Tracy, who needed someone to magnify their voices as their narrative was subsumed by the biases of others. I had legal perspective *and* personal empathy. Both were important.

Like the rest of the media, I had a press pass to get into the courtroom each day. Instinctively, as a former prosecutor I would bypass the entrance for journalists and the general public and head toward the queue reserved for court personnel. It was interesting because the law enforcement officers who protected the courthouse and screened visitors wouldn't redirect me. Instead, they let me in the same way they did the clerks, stenographers, and lawyers participating in the trial.

I knew without being told what to take off and put in the bins before passing through the magnetometer. And my friendships and good working relationships with officers and court personnel all those years I'd spent in Baltimore and Washington D.C. helped me to develop an easy rapport with this courthouse's team. They would tell me when the proceedings were going to start late and other tidbits I doubted they shared with other reporters.

While I tried to be diligent in my reporting and analysis of the proceedings, I began to see the same slanted perspective so prevalent at other media outlets creep onto the airwaves of CNN as well. For instance, the network hired an attorney who was vociferously and unapologetically on the side of the defense as a contributor. Other talking heads who would pop up fairly regularly echoed similar views. It was starting to be uncomfortably clear that CNN was trying to hype this distorted image of Trayvon, making him out to be an aggressive black boy who somehow provoked the death sentence Zimmerman felt he had the right to mete out.

It all made me bristle, and I began to counterpunch, offering more of my perspective on the air. I brought up racial disparities in the way the Stand Your Ground law was applied, how whites who attacked people of color got a pass, while African Americans didn't get the same leniency. I mentioned the details that were coming out on Zimmerman's background that the prosecutors brought up in court.

People noticed. I'd had the occasional person approach me for an autograph before, but now, I was frequently stopped when I walked down the street. Or men and women would approach me outside the courthouse and ask to take a picture with me.

"You go, Sis!" people would yell out as I headed to my car. "You're speaking up for us! We're proud of you!"

The attention surprised me. When you're in the moment, rushing to court, furiously taking notes, putting together talking points for when you appear on air, you're not focusing much on the larger world around you. But it became clear that this story was big, and yes, the world was watching.

Dad was watching too. And he was getting worried. "I saw you this morning," he said to me one day, catching me during a rare break. "You were getting a little fired up."

Unlike Mom, who'd once wanted to join the Black Panthers, Dad had been on the side of trying to change systems from the inside. He felt that by working hard and showing all that you were capable of, you could transform hearts and minds. Then with the goodwill and credibility you'd accrued, you'd be listened to when you observed injustice and spoke up for what was right.

"Are you trying to lose your job?" he asked me. He was serious. And the job wasn't his only concern. "People feel very strongly about this case. Maybe you're crossing the line. People in my office are talking about this. Especially my white colleagues. Are you safe down there?"

No doubt, the situation was tense. The trial, and all the issues it electrified, were creating factions across the country. There were daily protests outside the courthouse. People would drive by as we did live shots with Confederate flags flying from their cars.

One time we thought we heard gunshots, though it apparently was just a faulty muffler. But the rat-a-tat sound was enough to make us jump, our nerves were so taut. The atmosphere felt unsafe.

CNN actually hired security for the cameramen, on-site producers, and reporters. They would stay close, sitting in trucks or standing nearby when we were in court or doing a stand-up. And the members of our team would check on one another. The soundman would warn me not to stay out too late, and one of the producers would often walk me to and from my rental car along with one of the hired security guards.

Manny and I would talk every day, and he also grew concerned when he heard security had been hired. He asked if he needed to come down to Florida. But I told him I was good.

Dad had taught me to be disciplined, to reorient the system by being an inside player, but what he wasn't focusing on in that fraught moment was that he'd also taught me something else. When I decided to leave behind a full-time legal career and focus on television journalism instead, he became one of the loudest voices in the chorus, encouraging me to reach for the stars. He knew what it was like to yearn for something and forever wonder "what if." He told me that if you take the path that you're meant to follow, if you discover and embrace what you are truly meant to do, you'll feel like you never worked a day in your life, because your labor will be rich with meaning and purpose.

I felt that I had found mine. I wasn't going to stop speaking up about the injustice I was seeing, not just in the courtroom but among many of my colleagues covering this story. I wasn't turning back. Not even when CNN's response to my having all that access, all that insight, all that understanding, was to take me off the story.

I got that news in a phone call.

"Umm, hey, Sunny." It was the assignment desk I'd worked with off and on during my time covering the trial. "Management's been talking. Folks are concerned you're getting a little too close to the story, and y'know, we really can't have that. We need to be objective, fair, and not make the viewers think that we've got our own agenda. So, we'd like you to come back to New York. We're going to send somebody else down there to handle the rest of the trial."

I was shocked. Besides the fact that the network was featuring plenty of voices who were clearly offering perspective from the defense's point of view, I was the one who'd built this story from day one, when no one had even been interested in it. I was well-sourced within the courthouse, I had a rapport with Trayvon's family, and Ben Crump who was representing them, and even the prosecutor. And as an attorney with a journalism degree, I knew the law like the back of my hand, as well as how to convey complicated information in a way that the men and women on the street could understand.

If you want someone to watch and report on a surgical procedure, would you send in Sanjay Gupta or a general assignment reporter?

Of course it would be Sanjay. I felt no one at CNN knew their way around a courtroom better than me.

The network's decision was odd for another reason. I'm Catholic and made a point of saying so, yet the executives never had a problem with my talking about the sexual abuse scandal rocking the Catholic Church and the many legal claims being filed by survivors. I had prosecuted sexual abuse cases, never losing even one in court, and yet no one seemed to doubt that I could be objective when talking about a legal matter involving sexual harassment or assault. But a black boy gets killed, and now all of a sudden I was too close?

I felt that there were a couple of things going on. The story CNN had once dismissed as local was now one of the biggest headlines in the country, the kind of assignment that could make a favored reporter's career and reputation, and the network probably wanted to give someone else a chance to shine.

The other reality was that people of color are often accused of being "too close" or too sensitive when we deal with matters having to do with race. It's funny how that works. A white man can talk about anything—gender bias or bigotry, politics or business—and he's considered qualified to weigh in. But our actions are scrutinized whatever we do, and most especially, most ironically, when they pertain to issues for which we have a particularly astute perspective.

I had a decision to make. CNN wanted me back in New York, and I was on their payroll as a contributor. The safe, probably more sensible action to take would be to do as I was told. But I didn't. I decided that I wasn't going, even if I had to pay for my own hotel room.

CNN sent my replacement, a white reporter, immediately, in time for the next day of trial. The major media outlets each had an assigned space in the packed courtroom to guarantee them a seat. I had to give the CNN pass to the reporter who was now taking over, which meant that I no longer would be able to secure a spot. But I quickly came up with a plan.

BET, a TV network focused on African American programming, and Univision, a Spanish language network, had seats in the court-room assigned to them, but they didn't have reporters there con-sistently because it was expensive to keep a journalist on the scene indefinitely. I approached a couple of their staffers and asked if I could use their passes. They gave them to me without question. I will never forget those gestures. In fact, I kept those passes as a reminder of the solidarity that can be found among reporters in our industry.

CNN had begun splitting up its coverage, getting reports on the day's events from the correspondent on the scene, and analysis from a rotating group of local white male attorneys the network began fea-turing on air. I remember the CNN staffers who were rotated in and out being bewildered, then more than a bit irritated, when they saw me back in the courtroom, as if to say, "What is *she* doing here?" I might have asked them the same question. I'd stare right back, then not give them another thought as I put my head down and got back to work.

* * *

In the midst of the tense experience I was having covering the trial, I was dealing with an emotionally harrowing experience within my own family.

Even when I was busy, raising two kids and studying up on vari-ous legal issues so I could speak about them on TV, I made a point to head to the Lower East Side at least once a week to visit Nannie Virginia. She still lived there in that basement apartment with her husband, Tony.

She was so proud of me. When I entered the living room and gave her a hug, she would lean back and give me a broad grin. "There's my movie star," she would say, beaming.

Nannie was a voracious reader and consumer of the news. She would pore over every word in *El Diario,* a Spanish language newspaper, linger over her books of puzzles, and get her gossip fix reading the *National Enquirer.*

But I began to notice that the newspapers and puzzle books were stacking up on a table, going unread and untouched.

"What's going on, Nannie? You're not reading?" I asked. "Do you need new glasses?"

"No," she said distractedly. "It's too confusing to read now."

That was an odd thing for her to say. Nannie may have had only a sixth-grade education, but she was the smartest person I knew. She was aware of every vote coming up in city hall and the position held by every politician. Her ability to fix anything put in front of her showed that she had the mental acuity of an engineer. It didn't make sense that reading was becoming confusing.

In between my visits, I would call to chat with Nannie on the phone. She would ask me how the kids were doing, and I'd tell her about their swim lessons and sports games and what they were studying in school. She'd say good. Then increasingly, only a few minutes later, she'd ask, "How are Gabriel and Paloma doing?" as though I hadn't already told her.

My mother, my aunts, and I all began comparing notes.

Nannie had always been a vivid storyteller, and because she always had a new tale to spin, you'd have to make a request when you wanted a rerun of one of your favorites. But now it was like her mind was stuck on repeat.

At first, she would repeat the same stories she'd uttered a week before, during our last visit. Then she'd repeat what she'd said during our phone conversation the day before. Finally, she'd ask the same question or make the same point to you a couple of minutes after she'd already asked or said it. I was getting really worried.

Her husband, Tony, also began to complain that Nannie was frequently forgetting things. But it was when Nannie began to forget how to cook that I went into a full-fledged panic.

Nannie's meals filled a sweet spot in my soul. When I was away in college and coping with the sudden breakup of my parents' marriage, I would call Nannie, homesick. Instead of sending me a Tupperware container filled with one of her savory dishes, she would whip up a pot of *arroz con calamares*. Then, she would send the actual pot on a Greyhound bus and a separate pot of *carne guisado* and even sometimes send *alcapurias* ready to be fried. It somehow would get delivered to me. I would heat up the food in that very same pot. My friends always made it a point to stop by on the days they figured I'd be getting a taste of home from Nannie.

I inherited my love of cooking from her as well as from my mom. Frequently, when I'd be making dinner for my family, I would call Nannie for one of her recipes, though like most wonderful home cooks, she didn't have any written down, and her directions were, to put it mildly, less than specific.

To make rice and peas—*arroz con gandules*—for example, Nannie would say to take a cup—not a measuring cup but any glass you had in the house, and fill it with rice. Then she'd say, "Take the same cup and fill it with water. Make sure the rice in the pot is covered with the liquid, then add the *gandules*." You'd fill the emptied can with more water and pour that in as well. But just enough to cover it. More than that the rice would be too soggy.

But now when I called Nannie for her recipes, it wasn't to get a guide to how to make Christmas dinner. It was to test her memory.

"Asunción, I don't remember," she said more and more often, until the day came that she couldn't even recall how to make *arroz con gandules,* a dish that she once could have prepared while sleepwalking.

Not long after that alarming call, I went to Nannie's apartment. Tony was home.

Our family had never cared for him. It was understandable why he'd wanted to marry Nannie decades before, even though he was in his twenties and she was in her forties. She had been vivacious, beautiful, and sexy. But he'd never been worthy. He'd always held a good job at a factory, but he had a serious drinking problem. Nannie carried him, providing him with stability, and inevitably, after his few and fleeting attempts at sobriety, she would pick him up and dust him off when he fell off the wagon.

I walked into the kitchen and found all the knobs on the stove had been removed. Shocked, I stormed back into the living room where Tony was slouched, holding a beer. "What happened to the knobs?" I asked him.

"I can't trust her," he said dismissively. "She leaves things on the stove. And she's been causing fires."

I couldn't believe this was the first time I was hearing about this.

"You guys need to come here more," he continued, his voice starting to rise. "It's too much for me."

In retrospect, I understand that it's very difficult to be a caregiver. It was clear, and soon confirmed by doctors, that Nannie was suffering from dementia, and to care for someone going through that struggle requires incredible patience in the midst of your mourning for the person who is slipping away. But hearing those words at that time from Tony just made me angry. Someone in our family came by to see Nannie at least once a week. As for him, the vows were for better or worse, in sickness and in health. It was all great when she was forty and sexy, but now that she was eighty and sick, it was too much for him? That wasn't the deal, buddy.

Still, we knew better than to trust him entirely with Nannie's care. We were scattered at this point. My aunt Carmen and my cousin Jeffrey lived in New Jersey. Nannie's siblings were dead, and of course my and Jeffrey's children were young. There was a smattering of other cousins in Brooklyn, Manhattan, and Puerto Rico, but we were a small family that had to figure out how to rally in a hurry. We decided to hire someone to stay with Nannie during the day.

Since Tony would leave for work about eight in the morning, we asked the home health aide to come in about noon so she could make sure Nannie had lunch and stay with her until five, when Tony was due to be back home.

But the aide started calling my mother and aunts to tell them that it was her time to head home and Tony still wasn't back. We'd ask her if she could stay a little longer until one of us could get to the apartment. Sometimes, by the time one of us arrived, Tony was home. Other times he would amble in a couple of hours later.

It began to get worse. Sometimes Tony didn't come home at all. When my mom and her sisters get together, they're like the Witches of Eastwick, tough and unified. I joined their trio, asking Tony what the hell he was doing.

"I need a break," he complained. His wife of forty years was ill and he was staying away all night or sometimes for days. Was he serious? We told him he'd better get himself home. But our warnings fell on deaf ears.

The last straw happened one terrible weekend. By this point Nannie had slid into the early stages of Alzheimer's. I must have been traveling for work because I didn't make it over that Friday, and no one else in the family stopped by either. The aide left, probably assuming Tony or one of us would be there soon.

Tony didn't come home for the entire weekend, and at some point, my grandmother fell. She lay on the ground until Monday, when the home health aide returned and found her. Nannie had a horrible gash on her forearm, a cut to the bone so deep and infected that she needed to be hospitalized and to have surgery. She even required skin grafts.

We knew that she could not go back to that apartment. My aunt Inez suggested that we put Nannie in a nursing home, where she could get care and have someone constantly watch her while she recuperated.

I was uncomfortable with that. I felt that family should look after her the way she had always looked after us. But we all worked, making that difficult. I begrudgingly went along with Aunt Inez's plan.

We found a facility in Westchester, not too far from my home, where Mom lived with me, Manny, and the kids. Nannie's memory was fading, but she was lucid enough to know that her husband of forty years wasn't there. It hurt to see her confused, unhappy, and missing Tony.

"Where is he, where is he?" she would mumble in and out of the fog that was now constantly clouding her mind. She was heartbroken.

After Nannie's fall, Tony visited her only once, when she was still in the hospital recovering from surgery. He looked an absolute mess, like he'd been on a bender for days. It was the last time I ever laid eyes on him.

My mother and aunts had one more encounter with Tony. After

Nannie had left the apartment for what we knew would be the last time, they went there to get some of her things and also to lock Tony out. Nannie had paid the rent, working as a super in the building, and our family felt that Tony, who had been so irresponsible and then neglected Nannie to the point that she lay on a floor bleeding and hurt until an aide found her, didn't have any right to continue living in the apartment.

Tony arrived around the same time my mother and aunts did, and he obviously had a different opinion. He called the police, and because his name was on the lease and theirs weren't, they had to leave.

In the following days, we asked for some of Nannie's things, her jewelry, her favorite books, some photographs. And I asked for what was most precious to me: Nannie's cast iron pots. Those metal pans, scented and seasoned no matter how often you washed them with the garlic and adobo she used in her beloved recipes, were seasoned with Nannie's laughter, and redolent of the times I'd spent learning how to cook the food of Puerto Rico at her elbow. They were pieces of Nannie. And Tony wouldn't give them to me. My feelings for him curdled even more.

At the rehabilitation facility, Nannie began to complain that she was being ignored. I think that once the nurses recognized me from CNN, they started to treat her a bit better, checking on her more often, getting her an extra cup of Jell-O or juice. But I remained concerned about her complaints, the anecdotes I'd heard about such places mistreating patients, fixed in my mind. There also weren't many Spanish speakers on staff, which made the place even more isolating for Nannie. And finally, a bunch of nurses, doctors, and aides weren't family. It wasn't Nannie's home.

One night, I showed up there after working on *Anderson Cooper*

*360°.* I found Nannie, in the bed, covered in feces. Her hands were smudged with her own waste because she'd been trying to clean herself. I went ballistic, running to the nurses' station to ask how they could leave my grandmother in that condition.

Nannie had sciatica and arthritis, and so it was painful for her to move. The nurse said that they had tried to get her out of the bed, but Nannie had been combative. I didn't doubt it. Nannie was very feisty. She cursed like a sailor when she didn't want to be bothered, and she would have been even more fired up if she was experiencing discomfort. But I couldn't fathom how a bunch of so-called professionals couldn't handle an elderly, ill lady.

"So you let her crap on herself?" I said in disbelief. I turned and went back to Nannie.

The family had bought her these beautiful *batas,* or housecoats. I cleaned her, got her a set of fresh underwear and a robe, then called my mom. "I just came by the home," I told her. "And Nannie was lying in her own crap."

"Well, that's it," Mom said, uttering what I already knew. "She can't stay there anymore."

I went back to Nannie's room and asked if she wanted to leave the facility. She didn't hesitate to say yes. She said she wanted to go home, but it was time for a heart-to-heart. "You can't," I said. "Tony is a jerk. He left you alone for an entire weekend. You fell and hurt yourself badly. That's why you're here. None of us can watch you in your apartment. You'll have to stay with me and Mom."

But it wouldn't be that easy. I guess few things ever are.

You never understand the indignities and irrationality of our medical system until you have to deal with it, and of course by then you are at your most vulnerable, trying to get urgent, even life-and-death, care for yourself or your loved ones. I went to the main physician at the facility and told him his staff had neglected my grandmother and so I was taking her out of there. And he had the nerve to say no, that she couldn't just get up and leave.

I really think they just wanted the insurance money. Nannie had Medicare, and I suspect the home got paid a lot of money per bed. They fought us for at least a week, blocking us from taking her home, even though they weren't properly taking care of her.

By this point, my aunts, cousins, and I were rotating our visits, checking on Nannie as we tried to untangle the red tape. I guess it was in my DNA as an attorney, and as an honor student before that, to play by the rules. But it took Nannie to turn me into a rebel.

In one of her decreasing moments of clarity, Nannie confronted me. "I thought you said I was going home with you," she said.

"They said I can't just take you, Nannie. I'm working on it, but I can't break the rules."

The Nannie of old, who'd stomp down to the Hells Angels headquarters if they were making too much noise, and actually got them to quiet down, looked me straight in the eye.

"You need to break the rules," she said point-blank. I took that as an order.

I stuffed one of Nannie's robes and a few of her clothes into my big tote bag. Then I went to the person on duty at the desk and told her that I wanted to take Nannie out for fresh air. I got a wheelchair, took her outside, had her wait as I pulled my car up to the front door, then the two of us roared out of dodge like Bonnie and Clyde.

The facility didn't take what I'd done lightly. For a while they harassed us, calling and threatening to sue. But they were clearly just trying to intimidate us. It wasn't like they could hold somebody hostage, and they finally gave up.

Our getaway had felt like a sweet victory. I was driving a Mercedes then, and I remember Nannie nestling into the seat as I pulled away. "Oooh, fancy," she said, a satisfied smile creasing her still-beautiful face. I actually took a picture of her with my phone. I guess I knew it was a moment worth preserving, partly because it was so crazy and partly because it was a fleeting glimpse of the old, hell-raising Nannie. Her smug look of victory was magnificent.

I was right to capture it. My aunt Carmen, who is a registered dietician, found us another home health aide who was able to move into my home in Westchester, which was a godsend because at this point Nannie needed round-the-clock care. I wasn't blind to the fact that we were blessed to have the financial means to pay for such a service, unlike a lot of many other Americans.

But not more than six months later, Nannie had a heart attack. Manny, Mom, and I were all out at the time when Mom got a call from the aide, who said Nannie was experiencing terrible pain in her chest. After Mom called me, I had my nanny Suzy go to the guesthouse where Nannie stayed to check on her.

"We need to call nine-one-one," she said frantically.

When the ambulance arrived, I asked to talk to the emergency medical technician. He in turn asked me an odd question: Did I want them to take Nannie to Greenwich Hospital or to the medical center in White Plains?

Why, I thought, are they asking me where to take her? "Which is closer?" I asked. He told me White Plains. "Well," I said, "take her there."

I assumed that the closest hospital was the obvious answer. But I now believe that the paramedics took one look at my large home and, figuring that we were affluent, gave me the chance to redirect them to Greenwich, which was probably the hospital favored by the wealthy.

I don't know for sure that Nannie would have gotten better care there, but the question has haunted me ever since, almost as much as the realization that while people should be able to get the best, most necessary treatment in whatever hospital they are rushed to, that simply isn't the way this country's health system works. I beat myself up over that in-the-moment decision for years afterward. I still do, actually.

When I arrived at the hospital, I found Nannie not talking and exhausted, resting with her eyes closed. But I knew she would bounce back. In my mind, she was invincible. Then Manny came and had a conversation with the doctors. He told me that I'd better call my aunts.

The family gathered at the hospital, but after a while, we all left and went home to get some rest, change, and come back in the morning. It was all so surreal, I guess we just couldn't get our heads around the idea that there was any chance that Nannie might leave us. It was unfathomable.

My beloved nannie Virginia died of sepsis just a few hours later on October 15, 2013. She had been alone when she passed. To think of that causes me pain to this day. She was eighty-seven. At least we think she was. She frequently, conveniently forgot how old she was.

I was inconsolable, unable to get out of bed for several days. I lay there in the dark, drifting in and out of sleep, sobbing whenever I was awake. Of course, it would be Nannie who made me snap out of it, if not my grief, then at least my stupor.

I was huddled under the covers when I peered through the darkness toward the couch in my room. And I swear, I saw Nannie sitting there, her long dark hair flowing down her back. "Get up," I heard her tell me. "I'm okay."

Later that morning, my mother knocked on my door and entered my room. I told her the vision I'd had and how I didn't think it was a dream. She had been there. And Mom said Nannie had also come to her and said the same thing. "She wouldn't want to see you like this," my mother said gently.

We had the funeral at the Church of the Resurrection in the town of Rye.

The morning of the service, I of course was a mess. But my girl-friends rode to the rescue. Knowing that I was having a difficult time, they showed up at my front door from Maryland with several different dresses that I could choose from. They did my hair. They held me up.

We didn't invite Tony. We didn't even tell him Nannie had died. I don't think any of us trusted ourselves to see him because we still felt such anger at the pain he had caused Nannie in the last months of her life. Even now, as I recount what he did, how he behaved, my hands are shaking.

The service was beautiful. Gabriel did a reading, and the choir sang a few of Nannie's favorite soul-lifting hymns. She was cremated, and Mom kept the ashes.

But after Nannie's death, our family was never the same. She was truly the matriarch of our family, and I think you don't realize all that entails until that person is gone. She was not only the glue binding to-gether my aunts, my cousins, and me, she was the link to our culture.

When I was young, and even in the early years of having my own family, all the relatives would gather every Friday at Nannie's apart-ment to eat dinner and share stories. With her, we ate only Spanish food. And she was the last person in our family who exclusively spoke Spanish. My children were bilingual because when they addressed their great-grandmother, they had to speak in the tongue of Puerto Rico.

Trying to understand that and to do what I think Nannie would have wanted, I tried for years after her death to be the one to bring the family together, hosting dinners and celebrations. But recently, I've pretty much stopped trying. People don't show up. People say they are too busy.

Our matriarch, our glue, is gone. And I will forever miss Nannie Virginia.

* * *

The day finally came when the case of George Zimmerman went to the jury. As a prosecutor, I'd always been nervous waiting for a verdict, questioning if I had done my job well, wondering if I had done right by the sexually abused child I'd been fighting for. But the verdict in this case somehow felt even more weighty, like a sizzling wick that could either be extinguished or set aflame.

Would there be justice, an acknowledgment of the racist violence and brutality that constantly snuffed out too many innocent black lives? Or would those who already didn't believe in the legal system be vindicated in their doubts, anger, and grief?

It was the latter. The jury acquitted Zimmerman. He would go free. And I couldn't hold back. As soon as the trial was adjourned, I rose from my seat, almost trembling in shock, and when I saw straight over to Sybrina and Tracy, I said, "I. Am. So. Sorry." It was all I could say.

I think my family would have torn that courtroom apart. But again, Sybrina was the picture of grace. She did our community a service, representing us in such a terrible moment with more dignity than the broader society deserved.

I went on CNN shortly after court adjourned and expressed outrage. There was no being objective, no time for sugarcoating or beating around the bush. The prosecution had proven its case. The jury, perhaps because of its own biases in favor of defense attorneys, or against a teenage black boy, had just refused to see it.

I also knew the prosecutors had to be devastated. I had won all my cases as an attorney, so I had never known the pain of working so hard, painstakingly piecing together an argument, only to have it rejected. But I could imagine how it would feel. It's something that you never get over. I know Marcia Clark, the attorney who prosecuted O. J. Simpson in the infamous so-called trial of the century, which resulted in his acquittal for two killings, and to this day, Marcia's still affected by that case.

People may wonder whether I did get too close to the story. But rather than becoming too close, I believe that many others were too distant. One of the reasons I'd wanted to be a journalist was that I wanted to do as a reporter and legal analyst what I had done as an attorney, fighting for the underdog, speaking up for those who might not be able to speak for themselves. I was mature and professional enough to be able to understand and articulate the argument being made by Zimmerman's team, but it was critical to also point out how irrational and overzealous, and, yes, criminal, Zimmerman's actions were.

I had a perspective based on my lived experience that many other journalists did not have, as a black Latina woman who'd been followed around a department store, as a mother who could never think of my son walking around the world on his own in the same way. And that was valuable. That was needed. It gave me my purpose.

Mark Twain has many legendary quotes, full of humor and insight. But the one I like best may be "The two most important days in your life are the day you are born and the day you find out why." But I actually think he got that a bit wrong.

I think there are three defining days in your life. Besides day one, when you are born, and the day when you discover your purpose, I believe there is a day in between. And that is the moment when you find out who you will *not* be, who you will not become. Your day two.

Day one for me of course came in October 1968. My day two came seven years later.

The colors remain bright. I'm sitting on a cold black-and-white tiled floor, and there are splotches of red everywhere, rich and crimson, like spilled wine. My uncle Eddie James is splayed out beside me, stabbed and bleeding, and I am trying to sop up the redness, to push back the yellow of his intestines, as other adults appear, their screams and wails echoing off the bathroom walls.

I remember, in that moment, amid the yelling and the blood and the terror, being overtaken by a feeling of calm and conviction. "Uh-uh," I thought to myself. "I'm not going to live like this. I'm not going to become this. This is not who I will be."

It was the coda of my first years lived in the South Bronx, where I was cushioned by love but surrounded by drugs, alcohol, violence, and chaos.

If I had let my surroundings stunt my imagination, if I had let them shape my course, I would have been a statistic like my uncle Ed, who didn't die that day but did lose his life years later to AIDS, which he'd contracted after years of drug abuse. If I had let the environment that I was born into determine my destiny, I certainly wouldn't have gone to high school at twelve, to college at sixteen, and onto a successful career in law and television. Instead, those circumstances that I was born into defined who I would *not* be.

This is why, despite the criticism and questions, I, a black Latina, chose to become a prosecutor. I wanted to make sure those things I witnessed growing up didn't happen, and when they did, that those who committed such acts paid a price and never did them again.

Then, I met Manny and brought into the world my true gifts in life, Gabriel and Paloma, and I knew I didn't want to spend my days immersed in the psychosis and pathologies of sexual predators. But I still yearned to protect people. When I got the chance to be a legal analyst, I thought, yes, I'd do it. I could talk about these issues on television, translating my passion to a new platform. I was still fulfilling my purpose, just in a different way.

But I hadn't really found my voice. I hadn't really manifested my full self. I was cutting my hair and clipping my words to imitate Soledad O'Brien. I even sometimes tried to mirror the ebullience of my idol, Oprah Winfrey. "You get a car! You get a car!" It was silly.

I found my day three, my passion in life, in Sanford, Florida. I came into my own after that case and trial because I used my voice, without fear or hesitation, to speak about what I felt was right, to speak for someone who couldn't speak for himself. I finally began to fulfill my purpose because I became undeniably, authentically Sunny.

After Trayvon, my profile at CNN grew. I appeared on *Dr. Phil.* I even met Oprah. But more importantly, people would come up to me from then on, and while they didn't necessarily remember my comments on the Casey Anthony case, or the Catholic Church sex scandal, they did know me from my coverage of the murder of Trayvon Martin. They felt I'd spoken for them, that I made their fears, their rage, their humanity, matter.

In the years since, I've seen so many others find their day threes, many because of the wrenching, continuing cycle of killings of black boys and men, from twelve-year-old Tamir Rice, a child killed for playing with an air pellet gun, to Eric Garner, surrounded by a group of police officers and choked to death by Daniel Pantaleo for selling cigarettes on a New York street corner.

And then there was Michael Brown, killed by a cop in the St. Louis suburb of Ferguson, Missouri. For weeks, young people streamed into that city, peacefully protesting in the face of overaggressive policing. Do you know how cold it is in Missouri in the winter? And yet, they came. And yet they fought, peacefully.

I've seen many find their day twos, their what-not-to-bes, as well, whether it's a young person who videotapes and reports schoolmates hurling racist chants instead of joining in, or an adult challenging a relative who loves to make racist jokes around the holiday dinner table. Such actions are so much bolder, so much more courageous, than the Twitter thugs tapping out threats, drunk with their keyboard courage. I've come to know that group well.

In the years since I found my day three, since I began to speak loudly and often about the state of race relations in this country, I've been viciously attacked. I've been called a racist and a race baiter. And that vitriol has only intensified during the presidency of Donald Trump, whose rhetoric and administrative policies have been a clarion call to bigots, telling them that they have no reason to hide their prejudices, no need to silence their hatred.

But here's another thing you discover when you find your day three: You don't worry about the haters. Once you hear the sound of your own authentic voice, once you feel you've found a purpose that is more important than your own comfort, you can't ignore it. To me, a per-

fect day is trying to help eke out justice for those who are going un-
heard and doing something for someone else without any expectation
that they will do anything for me.

The Trayvon Martin story changed my life. Of all the cases I've
ever worked on, of all the stories that I've ever covered, none has impacted me more than his. Trayvon put in stark relief all the reasons
that I wanted to be a journalist. It solidified for me that all the ratings,
Emmys, and accolades in the world would mean nothing if I couldn't
use my position to spotlight stories some producer deemed to be too
small and to humanize people that some deemed unworthy.

Proverbs 31 says, "Open your mouth for the mute, for the rights of
all who are destitute." There are many people who dislike my outspo-
kenness. But to them, I say, bring it. We are all Trayvon.

# CHAPTER ELEVEN

# THE VIEW

In the years after the Trayvon Martin case, I was back at CNN, continuing to do analysis though I yearned to do more. I decided to go to the National Association of Black Journalists convention to network and attend workshops.

The NABJ job fair was crowded and filled with beautifully, sharply dressed black folks zipping around with résumés and reel tapes in tow to line up their next gigs or their big breaks. I went armed with my résumé and reel tape and met talent executives from all the major broadcast and cable networks.

I interviewed with NBC and CBS executives in New York, and although there was nothing for me right then, we kept in touch.

Not long after that producer told me that I didn't have what it took to be a national anchor, I got a call from ABC offering me a chance to be a substitute anchor for the overnight show.

*World News Now* and *America This Morning* came on at 2:30 A.M. and stayed on through the wee hours. But as late as it was, everyone in the industry watched it, as did over two million other viewers. (Insomniacs, new moms, shift workers, hospital staff, cops, a lot of people.) "Do you want to do it?" I was asked.

I was shocked. I'd never done that kind of show and had been told that I wasn't good enough to anchor even a legal program, firmly centered in my expertise. Now I was being asked to host a national show without even an audition.

My status as a contributor at CNN, as opposed to, say, being an official correspondent, gave me the flexibility to be able to take on outside gigs. Of course I said I would take the job.

But there was a definite learning curve, a rather steep one, if I'm being totally honest. I didn't know even the smallest, most basic things, like what the letters "fs," etched in the corner of the teleprompter, meant.

Luckily my co-anchor Rob Nelson, who is still a dear friend, was a patient teacher. "That's full screen," Rob would say calmly. And he didn't do it during the commercial breaks or as a whispered aside into my ear. He'd give me instruction right on air, not to embarrass me but to reassure me that there was no shame in being new and not knowing something.

Another time I was reading the teleprompter. I wasn't familiar with the Nikkei, the Japanese stock exchange. "And the Nicky average," I said, pronouncing the word like I was pronouncing the nickname of an old friend.

I heard the crew start giggling.

"Oh, did I mess that up too, guys?" I asked sheepishly.

"Yeah!," someone yelled from off camera. "But don't worry about it."

That's what was so great about the whole experience. Unlike those early days at CNN, when I felt humiliated and alone, Rob and the crew made me feel like I was in a safe space despite my multiple gaffes, like they were content to help me steer atop my training wheels until I was finally ready to ride by myself. It felt good.

With the help of Rob and so many others, I learned to anchor a national news show.

I actually was a "substitute" anchor for an entire year and a half. I'd head home from CNN, maybe take a short nap before having dinner with Manny and the kids, then drive to ABC's headquarters on the Upper West Side of Manhattan. I'd arrive at 10:00 P.M. and when I left the building eight hours later, the skyline would be painted a rosy gold as metal coffee carts clattered up to their spots on the battered curbs to wait on the morning rush.

It turned out that one of the rabid insomniac fans of *World News Now* and *America This Morning* was Whoopi Goldberg. Whoopi is one of the few people on Earth who's won an Emmy, a Grammy, an Oscar, and a Tony. After a career that included starring in *The Color Purple* and *Sister Act*, along with being the star of a groundbreaking one-woman show, and one of the most inventive hosts in the history of the Oscars, Whoopi became the new moderator of *The View* in 2007. Beyond her multiple talents, including a wicked sense of humor, Whoopi is incredibly well informed and thoughtful. She watches lots and lots of news.

I soon found out that after seeing me anchor one night, Whoopi mentioned me to Bill Geddie, who co-created *The View* with Barbara Walters and was its executive producer for nearly twenty years.

"I saw this girl Sunny Hostin, and she was really, really good," I learned she told Bill. "We could use someone like her on our show." Whoopi thought I was funny, and she saw the similarity between a segment Rob and I did called "The Mix" and "Hot Topics" on *The*

*View*. Both required rapidly diving into a mélange of the day's most-talked-about headlines, and Whoopi saw that I could handle it.

Of course, I didn't know any of this was happening. To me, the failures, flubs, and rejections were top of mind. That CNN producer who said I wasn't good enough to anchor my own show reverberated in my ears and weighed on my confidence. But literally a couple of months later, I was co-anchoring a national news show that everyone in the industry watched. And Whoopi Goldberg of all people was talking about me to the executive producer of *The View*. That meant a lot.

ABC hired Paula Faris to become the permanent *World News Now* and *America This Morning* anchor and I was given a wonderful good-bye tribute by my ABC overnight colleagues, even though I was never an official anchor. I again focused fully on CNN. But I believe my stint at ABC grabbed management's attention, making them see me in a different light. Things started happening for me.

First, I got a show with the famous criminal defense attorney and my good friend, Mark Geragos.

The network actually built Mark and me our own set, which was a pretty big deal. It meant that before we even got out of the gate, the executives thought we had a chance for a good long run. I remember that when Geraldine Moriba, the head of diversity for CNN, laid eyes on it, she was so ecstatic she literally ran down the stairs and hugged me so hard, she lifted me off my feet.

"You did it!" she yelled. She was the senior executive producer of Soledad O'Brien's *Black in America* series and worked incredibly hard to make sure the voices of people of color were heard and that we got to tell our communities' stories. She was truly a solider in the battle for equity and diversity. She was also a tremendously generous spirit who felt that whenever another person of color advanced, so did she. And she was a real advocate for me. I must say that despite my rocky start at CNN, I'd developed some truly special relationships with an amazing group of people, like Anderson Cooper, like Don Lemon, like Geraldine.

Finally, on a Friday in March 2014, Mark and I had our debut. Mark is a very calm, cool customer, but even he was excited. We knew each other well and really clicked on air. But that first show turned out to be our last. On March 8, 2014, Malaysia Airlines Flight 370 and the 239 people it was carrying from Kuala Lumpur to Beijing disappeared. Of course, the tragic mystery dominated the headlines as the largest search in aviation history ensued. Regular programming was preempted to focus on the investigation. Our show never got a second episode.

Amid the round-the-clock coverage, I was initially scheduled to fill in as an anchor from time to time. But again, the word no intervened when someone decided that I wasn't ready to handle the rolling, non-stop coverage of a breaking-news story. I was frustrated. But soon, I would be given another chance. A really big one.

Jeff Zucker had been deemed a media wunderkind. He was credited with sparking the then unprecedented success of *Today* and eventually became the president and CEO of NBC Universal. In 2012, he came to CNN and soon assumed the mantle of president. Now, roughly a year after being in the top job, he gave the green light to my having my own show.

I wanted to be able to have an opening monologue, giving my take on a legal issue of great importance to Americans, like reproductive freedoms or gun rights, as well as feature interviews with an array of interesting guests. For my pilot, I was able to discuss topics and debate with Alan Dershowitz, the well-known attorney who has recently been in the spotlight because of his defense of some of Donald Trump's controversial behaviors and policies and because of his alleged involvement in the Jeffrey Epstein scandal.

At the end of the episode, the show highlighted missing black women and girls, something that was very important to me. The show was well done, I believed, topical, urgent. But we didn't get the go-ahead from Jeff for a full run. I began to believe that a big breakthrough for me at CNN was never going to happen.

My last big opportunity came a short while later, when Jeff made me part of a new show that was going to be similar to *The View* as it had been envisioned when it first appeared on the air, with a group of women talking politics. There were four of us, the lawyer Mel Robbins, the liberal LGBT activist Sally Kohn, and Margaret Hoover, a conservative pundit who also happens to be the great-granddaughter of President Herbert Hoover. And then, me.

Our executive producer was an African American woman based in Atlanta who was well regarded in the newsroom and throughout the industry. This was a big opportunity for her, as well as for each of us who'd been chosen as co-hosts. We were on for several weeks in what amounted to a lengthy tryout to see how we vibed with one another and how the audience responded. Things were chugging along. And then in maybe the fourth or fifth week, we made a fatal error.

It was in the midst of the protests growing in response to the killing of Michael Brown. Activists were raising their arms to protest not only Brown's killing but also the constant shootings in which police officers killed black boys and men even when they raised their hands to show they didn't have a weapon and weren't resisting. Sally felt we needed to make a statement and suggested at the end of this particular show, we all raise our hands in solidarity with the demonstrators and others who were sick and tired of these outrages.

I had a slightly different idea. I was riveted and sickened by the case of Eric Garner, who heartbreakingly, repeatedly said, "I can't breathe," as a horde of officers wrestled him to the ground. All of us on the show agreed that the violence inflicted by police on African Americans was an epidemic, and we needed to say something. It was decided that we could each respond in the way we saw fit. We didn't, however, clue in our executive producer.

I wrote the words "I can't breathe" on a sheet of paper. At the end of the show, we stuck to our plan. I held up that paper, while the other three hosts held up their arms as if to say "Don't shoot." The cameras went dark. And Jeff Zucker just about lost his mind.

Apparently our show had been doing a lot better than I realized, and I believe Jeff was on the verge of greenlighting it for a full run. But of course, after our stunt, conservative commentators had a field day, and the complaints about our taking a stance poured in.

We tried to contain the damage. We needed someone to go in and apologize to Jeff and hopefully save our show.

"Well," Sally said, "I don't think the six-foot lesbian should go to him. I doubt he'll be very sympathetic."

"Well," I chimed in, "I don't think the black woman who refused to come back from Sanford, Florida, when she was ordered to should plead our case."

Clearly, despite our predicament, we hadn't completely lost our sense of humor—or our minds. Finally, we all agreed that the person descended from the thirty-first president of the United States should be the one to ask for mercy.

It didn't work. And honestly, Jeff was right not to give us a second chance. I'm normally such a rule follower, but in the moment, I didn't think about the many people who would be impacted by our decision, not the least of which was the African American woman whose chances to ever again be an executive producer we probably ruined. You don't go rogue like that in television. I would never do it again.

But at that point, I'm sure Jeff wasn't interested in hearing my regrets or about the lessons I'd now learned. He probably thought, "I've given this chick three chances. Three strikes, you're out."

Around that time, *The View* came calling.

\* \* \*

I'd been dancing with the show for a while.

After Whoopi brought me up to Bill Geddie, I began being invited to fill in on the show. Those pseudo tryouts had begun a couple of years before, in 2012. I developed great relationships with some of the other co-hosts, like the comedian and actress Sherri Shepherd and even the grande dame herself, Barbara Walters. When I sat on the set, sparring about politics and pop culture, I felt incredibly comfortable. I felt like, "I know how to do this." It felt like home.

At one point, the show's executives even asked me to officially become a contributor. Others like Padma Lakshmi, the co-host of *Top Chef,* had also been filling in as an every-so-often host. But I passed. I'd been a longtime contributor at CNN, and though I was treated in many ways like a full-time staffer, with an office, a company credit card, and some other perks, I still felt that status kept me on a lower tier. I didn't feel like stepping into that secondary status somewhere else.

But Whoopi kept on talking about me. And one day, I got an out-of-the-blue call, asking me to audition to be one of the permanent co-hosts of *The View.*

That was music to my ears. But if I got the job, I would no longer be able to contribute to CNN, so I felt that I owed it to Jeff to give him a heads-up about the potential opportunity.

I told him I'd been invited to audition. "With your permission, I'd like to go for it."

Jeff gave me his blessing, as well as a dig at the same time. "Sure, that's fine. But if I were you, I wouldn't count on getting it."

Once again someone was telling me what I couldn't do. And that just made me want it more. I'd learned what to do when I really wanted something. Megyn Kelly had taught me well.

* * *

Since I'd started in television, the message that I needed, to stand out, that I should never hold back when I had the chance to shine, kept coming up. I remember Don Lemon telling me when we grabbed lunch or coffee, "You're good Sunny, really good. But you've got to lean in."

My memory of Megyn dominating our segments on *The O'Reilly Factor*, and the constant advice from people like Don, stayed on my mind. When the right moment came, I had to recognize it and grab ahold of it, because it might never come again.

Bill Geddie seemed eager to work with me. He'd even told me after one of my times filling in, "Sunny, you've got to stand out. You've got to have a moment."

There was that advice again. Stand out. Lean in. Grab your moment.

I was ready. But the timing had to be right. After some of the original *View* cast members like Star Jones and Meredith Vieira moved on, *The View* tended to switch out hosts every couple of years or so.

I began hearing rumors that it was on the verge of making another change. During that time, Bill called me once again. He said he had an idea for a segment called "Ask the Lawyers," and would I come in for a test run of the concept?

Since I already gave my legal take when I appeared on *The View*, I figured that maybe I would be paired with another attorney, like I often was on CNN, and maybe Bill just wanted to get an idea of what pair had the right chemistry. But when I arrived that day at the studio, I found that there were two other attorneys there. Both of them light-skinned, both of them with a television profile. One, Shawn Holley, had represented Lindsay Lohan. The other, Loni Coombs, was a frequent TV legal analyst. Each of us looked surprised to see the other and the wheels in my head started turning.

"They're looking for another lawyer like Star Jones," I thought. Bill had called in three black women, all of us who fit a certain type. Forget a segment; this was a tryout for a co-hosting spot.

This was it. Time to lean in.

I got on that set, and those other lawyers never saw what hit them. I laughed, I argued, I looked straight into the camera. They couldn't get a word in. They looked thunderstruck. I Megyn Kelly'd them.

When we were done, I walked past Bill. I didn't ask how I'd done. I didn't have to.

"Good girl," was all he said. I never saw Shawn or Loni on *The View* set again.

When I finally got called in for the official audition, I wanted something new to wear but didn't have time to get to the mall. ABC's wardrobe department, stocked with flattering outfits for its anchors, actually agreed to lend me a beautiful dress. I picked it up, changed quickly, and headed over to *The View* set for what I figured was basically a perfunctory, obligatory tryout. After all, I'd hit it out of the park during that pseudo tryout a few weeks before. I knew everyone on the show and often filled in. I felt confident.

Imagine my shock when I showed up and saw something like a dozen or more women pacing around.

We were stuffed inside a green room. There was Stephanie Ruhle, who had worked in finance and been a correspondent for Bloomberg. Nicole Wallace, the former Republican campaign strategist who did eventually land a co-hosting role on *The View* for a year before going to MSNBC, was also there. I saw Sage Steele, an African American woman who's at ESPN; commentator S. E. Cupp, whom I'd worked with at CNN; a woman named October Gonzalez; and Lauren Sanchez, a television personality. And others.

Inside the studio, there was a round table. Whoopi was there, and so was Rosie O'Donnell, who had left the show but apparently was back.

I wondered what the hell was going on. I'd thought the job was mine. And I don't think I was alone. As I looked around the crowded green room, everybody looked out of sorts and uncomfortable. We were all dressed to the nines, and it became apparent really quickly that we were all vying for a spot on the show. It was just a big cattle call.

Each of us was given a list of hot topics. Producers then grouped us according to the topics that most piqued our interest.

There was definitely a Latina contingent there, indicating that a Latina voice might be something they were looking for. I had already encountered some of the same old, tiresome questioning when I'd mentioned on set that I was Latina, and once we got off air, people would ask how that was possible. The two women I noticed, Lauren Sanchez and October Gonzalez, definitely physically fit more of the stereotypical idea of what a Latina looked like.

Looking over the hot topics, October mentioned that there were a lot of issues dealing with politics. I explained to her that at one time, politics was what *The View* was known for. They'd moved away from it, but now I guess they were returning to their roots.

"Oh," October said as if she was trying to stifle a yawn. "I've never voted." She also mentioned that she didn't speak Spanish because she wasn't in fact Latina. The last name apparently came from someone she was not really married to, Tony Gonzalez, a former NFL player, who had been married to Lauren Sanchez, who I believe was in fact Latina, but not thrilled that October not really married Gonzalez was there. You really can't make this stuff up.

The whole vibe of the room was bizarre. Between that and the cacophony of conversations going in that green room, which reeked of perfume and sweat, things were clearly going to go very badly. What exactly was I doing here?

Finally, I was paired with S.E. Cupp, my fellow CNN contributor, who happened to be several months pregnant. We were going to be on set with Rosie O'Donnell, talking about abortion and birth control.

S.E. was a conservative and a gun enthusiast. I also knew that S.E. was an atheist, though I thought since she was now about to have a child, her view on spirituality might have changed. What she didn't know about me was that I was Catholic and pro-life. Given my perceived progressive politics, I figured that might take her by surprise. I expected that we would be able to engage in a lively discussion.

We went out on set and found Nicole Wallace at the table, along with Whoopi and Rosie. I'd mulled over some points about faith balanced with the right to individual choice that I wanted to make, but I needn't have bothered. Rosie and S.E. kicked off our conversation about abortion, and it immediately exploded into a heated argument. I couldn't get a word in. As the two of them went at it, I looked at S.E. and switched my mind-set from that of a contestant on this insane game show to that of a concerned mother. "This situation can't be good for a pregnant woman," I thought.

"Have you ever had an abortion?" Rosie yelled, stretching her neck toward S.E.

"That's none of your business," S.E. spat back.

Rosie, a former co-host who knew *The View*'s playbook, punched back at S.E. like a boxer who'd been on the ropes and suddenly got a second wind. "On *this* show, you talk about your business."

With that, the fake segment was over. We all headed to the green room, exhausted from the fight, even though half of us hadn't even been in it. S.E. sat on the floor next to her bag. We hadn't caught our breath before Rosie came storming in. She wasn't finished fighting.

She and S.E. exchanged a few more choice words, then Rosie left.

"I guess I'm not getting this job," S.E. kind of mumbled under her breath. I noticed she looked a bit flushed. We weren't exactly friends, but we were colleagues at CNN, and mother to mother-to-be, I felt I needed to look out for her. I asked if she was all right. "Yeah," she said, fatigue in her voice. She seemed defeated and angry. I left her alone.

The whole day was like a recurring nightmare you can't wake up from. At this point, people were dropping like flies, and I was one of the last people standing, called out to debate with a rotating group of hosts multiple times. Nicole Wallace was hanging pretty tough as well.

I was brought back out one more time after the Rosie/S.E. Cupp battle. I was in a group that included October "I've never been in a voting booth" Gonzalez. She didn't have much to say, and the discussion in contrast to what I'd just experienced earlier was pretty calm. Still, I was glad when it was over.

Soon I got news that all that madness hadn't been in vain. The network asked me to sign a deal sheet, which was basically a document stating that I wasn't going to take on another opportunity that would prevent me from joining the show. My agent said that was a great sign. And I definitely still wanted the job, even though the audition had been brutal.

When I got back to CNN and ran into Jeff, he asked me how things had gone, and I told him that it seemed very competitive. I didn't tell him, however, that I was feeling pretty confident that the job might be mine.

Leaks started appearing in the press that said I was in the lead to become *The View*'s next co-host. My agent and I kept in close touch.

"I'm told they haven't made a final decision," he said, "but it's pretty clear that you're probably going to get it."

While I was on pins and needles waiting for confirmation from *The View*, my contract at CNN was also reaching its end, so it was time for me to renegotiate. I needed to know where I was going to be, at CNN or ABC. The clock was ticking, but not fast enough for me. I wanted *The View*. I needed an answer.

I remember attending the U.S. Open, basking in the sunshine, having a lovely day. People approached me constantly, both acquaintances and strangers, congratulating me on the new job. So did many of my colleagues at CNN. They'd read the headlines; they'd heard the industry gossip. "This job is mine," I kept thinking. I even mentioned to my producers that *The View* would be starting to tape very soon, and there was a chance that I might be moving over there. I felt it. I believed it. I was ready.

I was at the Open, drinking champagne, literally celebrating my new gig. Then the phone rang. It was my agent.

"Hi, Sunny," he said. I could hear it in his voice. I knew. "By all accounts this was supposed to be your job, but they decided to go in a different direction."

"But they had me sign a deal sheet," I said, not quite believing what I was hearing. "If not me, who got it?"

He waited a beat. "Rosie Perez. They really wanted a Latina for the seat."

"But I am a Latina!" I practically shouted. It didn't matter. They'd made their decision.

It was a gut punch. My whole life I'd been told I wasn't Latina enough. And here it was again. I'd been a successful guest host. The tabloids were filled with headlines about my being the front-runner. People were congratulating me. And I lost out because the powers that be deemed that I, a Spanish-speaking woman who was Puerto Rican, wasn't Latina enough. It felt like the culmination of every doubt, every suspicious glance, every round of "How Puerto Rican are you?" I'd had to deal with my whole life.

*Do you speak Spanish?*

*What food do you eat?*

*What town on the island are your people from?*

All those questions, which had haunted me, infuriated me, saddened me, bubbled up in my mind, spinning on an endless loop. They collided with the other statements that stoked my self-doubt: "Stay in your lane." "You don't have it in you to be a national anchor." "Go ahead and try out. But I wouldn't count on it." I was crushed.

And now I needed to go back to CNN and tell them I was ready to renegotiate my contract after all.

I called Jeff, letting him know that I didn't get the job. I signed a new deal with CNN, and to add insult to injury, it was for less money than I'd earned before, which I'm sure was intentional.

It was like a funeral procession parading by my office those first few days. Some people wouldn't look me in the eyes, unsure what to say. Others, meanwhile, seemed to revel in my misery, only too glad to bring up what happened.

"I heard about *The View*. Soooooo sorry." They dragged out their words, the expression on their faces more smug than sympathetic. Thank goodness for Don Lemon, who had an office across from mine.

"Keep your head up," he said. "We've all lost jobs."

So I just went on, which is what we all have to do, in our careers, in our lives. I still had a job. I still had a platform. I still had a voice. I had to quell my disappointment and fix my eye on all that I had, not the loss of what was never truly mine. But I can't lie. It was very difficult to go in to CNN every day. And I stopped watching *The View*.

As it so happened, that turned out to be a challenging season on *The View*. The ratings dipped, Rosie O'Donnell left after only a few months, and Rosie Perez also left. So in a way, I suppose it was a blessing that I didn't get the job. I probably wouldn't have lasted, pushed out so they could rebuild the franchise.

It was a year or two later, maybe less, that ABC contacted me, asking me to come back. What's the old adage? Fool me once, shame on you; fool me twice, shame on me? I wasn't going back into that vortex again.

"No freaking way," I said in not so many words.

But I also knew that I needed a change. That I would not be able to tap into my full potential at CNN.

I was offered the opportunity to be a legal correspondent for ABC News, making appearances on *Good Morning America,* which had taken the crown from *Today* to become the number one morning show, and for the news division as a whole. I'd also be able to fill in on *The View* when one of the co-hosts was away.

In many ways, it was a dream job. Bill Geddie, who I really respected, was unfortunately no longer helming *The View.* But Whoopi was still there. And the comedian Joy Behar, one of *The View's* original co-hosts, who I'd met and adored, was rejoining the show. I'd have a strong support system at ABC, I thought, people in power and around me who believed in me and my talent.

I met with *The View's* new executive team, Brian Teta, a former David Letterman producer; Hilary McLoughlin, past president of Telepictures, a studio that I was actually talking to about a possible joint project; and Candi Carter, who had been a producer for the oracle herself, Oprah Winfrey. It was feeling like the right thing to do, like all the pieces were coming together.

I wasn't promised a permanent perch at *The View's* table. It was more like "We're rebuilding. Come on over, be a correspondent, and we'll see where it goes."

I'd wanted to be a reporter and to move away from analysis for a while. And I'd get to be a host on *The View,* albeit as a guest. Honestly, after hearing about the rocky period the show had been going through and in the wake of the traumatic experience I'd had during my last tryout, an occasional hosting gig suited me fine. I was a bit gun-shy and preferred to test the waters there before diving in.

I finally tuned into the show again, watching it a few times to get an idea about its pacing, its topics, its chemistry. I knew that I could do the job. But I still felt unsure. Being told what I couldn't do at CNN, the pilots that went nowhere, the previous job at ABC that I thought I would get only to be told I was apparently not Latina enough, had all taken a toll in ways I didn't always want to admit.

So I went to see Charlie Moore, Anderson Cooper's executive producer. He'd always been someone I could talk to, and had been a supporter during my coverage of Trayvon Martin's murder. Charlie had also been my executive producer on the short-lived show I'd done with Mark Geragos.

I told him about the offer from ABC. "But I'm not sure," I said. "Maybe TV isn't the best use of what I have to offer. I've also been thinking about maybe going back to the law full time, working for the District Attorney's Office in Brooklyn. Maybe I can make more of a difference that way."

Charlie was sitting in the control room at *Anderson Cooper 360°*. He leaned back in his chair. "You have such a big voice," he said. "You can reach way more people staying in TV than working in a DA's office. And at *The View*, the audience on a regular day has three to four times the viewership of any one of our shows. I hate to lose you, but this a no-brainer. You've got to take it." That was Charlie. Honest, always there for me, and, as usual, making so much sense.

The ABC package was a chance worth taking, an opportunity too good to pass up. I needed to go into ABC with no expectations, but with the confidence I could handle whatever presented itself.

For what would be the last time, I went to see Jeff.

I told him what ABC had offered me, that it was a job I had to take. I was immensely thankful for all the opportunities CNN had given me, but I had to give my two weeks' notice.

"That won't be necessary," Jeff said coolly. "You can leave today."

I'd like to say I wasn't expecting that response, but I must have sensed what Jeff would say because I'd already packed up my office before going to see him. By the time I walked the single flight of stairs leading to my office, the human resources department was ringing my phone asking me to turn in my building identification and corporate credit card.

Jeff's reaction could be read as cavalier, even mean. But the truth was, I don't think he'd ever been that convinced of my talents. He'd kind of inherited me from the network's previous president, Jon Klein. I've seen him since, and he's congratulated me on my success on *The View*, but I still don't think he gets me. And that's okay. Everyone has their preference.

But to his credit, though I was never one of his favorites, Jeff gave me chances, several of them. It just so happened that none of those opportunities worked out. As a woman of faith, I don't believe those were coincidences. Those shows were just not meant to be.

CNN had marked my big break in television, where I learned the ropes, gained confidence, and got a necessary, realistic look at the world of TV news. I got a few battle scars along the way, but they served me well, giving me experience and toughening me up. In a sense, I became the OG, the original gangster among African American contributors at the network. I feel grateful that I was able to forge a path.

Joey Jackson, an attorney who now frequently appears on CNN, has called me for advice. Angela Rye, a laser-sharp, blunt-talking attorney and political commentator has gotten in touch during her contract negotiations, and I've freely shared my salary history. Same for April Ryan. I've done the same with Bakari Sellers, the former South Carolina state senator who is also now a CNN contributor. I'm happy to share whatever wisdom and insights I can. CNN gave me a lot, and I am happy to pass the baton.

That day, I picked up my couple of boxes, turned in my cards, and walked out of CNN for good. Next stop: ABC.

* * *

The idea was that I would be ABCs legal correspondent, but I would fill in when the official co-hosts were away. But I think the producers soon realized as we went along that I did the work. That I was the one who showed up without complaint and could always be depended on.

*The View* also has an audience that is disproportionately African American and Latino, and I'd become very popular with a lot of our viewers. Little by little, I began to appear more and more. There were people with bigger names, like *Fuller House* actress Candace Cameron Bure, or Raven-Symonè, but finally there was consensus that I was really good at this.

I became a full-time co-host of *The View* in 2016. But I never got a proper introduction. I just went from being at the table a whole lot to being there every day. Unlike other co-hosts, from Rosie O'Donnell to Jenny McCarthy, who heard the words "Welcome the newest host of *The View*!" I never got such fanfare. There was not even an official announcement in the press.

Today, I am the host with the third-longest run of any on the current panel. I've been there longer than any of the crop of recent hosts, like Raven-Symonè, Paula Faris, Meghan McCain, and Abby Huntsman.

Yet in the fall of 2019, after I'd been an official co-host for four seasons, I'd still run into people who weren't sure if I was a permanent member of *The View*. "Did you become a co-host?" they'd ask innocently. "It's like you're there every day."

It would never have been a question if there had been a proper announcement, instead of the producers quietly switching my title on the show's website one day from "contributing co-host" to simply "co-host."

Frankly, it infuriates me. I think it's disrespectful. I've asked why I didn't get the big callout and told my producers that I'm still waiting for it. Their response has been, "Well, why? After all, you're here." But that kind of acknowledgment matters.

The lack of a formal introduction was indicative of a broader pattern, I believe. There were other examples of my being treated like a stepchild, like when it came to our dressing rooms, something that may seem trivial but symbolically means a great deal.

The permanent hosts had their dressing rooms on the second floor, which was where the action happened, from the food that was brought in, to the production meetings, to the wardrobe room. The guest dressing rooms were upstairs on the third floor.

When they were transitioning out Candace Cameron Bure, Jedediah Bila and I were the main guest co-hosts filling in, getting a calendar each month with the alternating days that we would appear. Jedediah was a Fox News contributor who brought a libertarian voice to the show. When Candace left for good, Jed was moved downstairs with the main hosts, into Candace's old dressing room.

Meanwhile, I remained upstairs in no-man's-land, prepping, as everyone else had breakfast. The producers would sometimes have to come find me.

Jed and I were and still are friends, but I felt dismissed and devalued that after my long history with the show, someone so new on the scene was suddenly appearing just as much as me, and now, inexplicably, she had gotten a dressing room on the main floor. In hindsight, I think that I was still having to prove myself, competing with a newcomer after I'd already shown what I could do. It was clear that they felt her voice was more valuable.

I asked Candi Carter if they were going to move me to a new dressing area. She said yes, but nothing changed. Then, when I brought it up again, reminding her that they'd moved Jed, Candi's surprising response was that they were thinking of having me share a dressing room with Raven-Symonè because next to Whoopi, she had the biggest one.

Well, I mentioned that to Raven, and you can imagine her reaction. No star wants to share her dressing room. She wasn't having that, and I didn't blame her.

I was moved eventually downstairs into a dressing room on the floor with the other hosts. Raven's dressing room went to Joy Behar, which was fitting since it was the second biggest, and along with Whoopi, Joy was an esteemed veteran of the show. My dressing room, as it so happened, was the third largest.

Again, it might seem like a small thing, but symbols matter. I'm no diva. I have the option of having a personal driver, but I refuse it because it's just not my style. I wouldn't mind giving someone a job, but the idea of being chauffeured around makes me uncomfortable. I also like the idea of having the freedom to come and go whenever I please.

But having a dressing room on the main floor, I believe, connotes how much you're valued, just like an official welcome and introduction to the millions who watch *The View*. Not having either made me feel like a second-class citizen.

It's ironic, because many now say that I'm the voice of reason on the show. The executives often comment that after a lot of changes and unrest, I'm steady. And some of the younger members of the team comment that the reason I'm not parodied on *Saturday Night Live* is because there's nothing to make fun of. I'm the normal one, they say, like the character of Marilyn on the 1960s sitcom *The Munsters*. I do feel now that I'm shown some respect, but I think the road it took to get there was longer, more winding, than it needed to be. And I still can't quite figure out why.

\* \* \*

There are those who are valued because of their last names or their connections rather than the dues they have paid or the skills they have honed. If you work the longest, if you work the hardest, that should count for something, whether it means getting paid more, garnering recognition, or simply being able to get a second-floor dressing room without having to ask for it.

Don't get me wrong. I don't have a problem with people having advantages. My kids, while they will have to deal with challenges because they are people of color, certainly have more than others. But I think privilege has to be acknowledged. You can't be born on third base and pretend that your home runs stem from the same kind of effort that someone born on first base had to expend. Some people don't have a mitt. Others don't have a bat. That needs to be recognized if we are ever going to resolve the many inequities that still dog our society.

But most days, rather than focusing on those broader dynamics, I'm just trying to prepare for my very busy days at work. They start early. If I'm going to appear on *Good Morning America*, I'm up at 4:30, but I can get an extra hour of sleep if I'm going straight to *The View*. I spend some time relaxing, feeding the dogs, and tending to my hens. Then I wake my kids to tell them that I love them, and I'm usually in the car by 6:45 and no later than 7:00 A.M.

I drop my mother off at her job at a daycare center, relishing the time we have to chitchat or just glide down the Henry Hudson with the radio purring in the background. I usually listen to the news on Sirius. It might be the *Today* show, one of the cable news shows, or an ABC podcast that sums up the day's headlines.

An hour later, I'm at the studio, in time for our meeting on the day's hot topics at 8:30. The previous evening, each of us will have gotten a list of about sixty headlines with hyperlinks to stories about politics and what's happening in entertainment. Because I'm such a news junkie, I'm typically already familiar with about half of them, and there are times I email my own suggestions to the producers. Whoopi and Joy often do that too.

Needless to say, we all have different stories that catch our attention. They might range from Donald Trump's impeachment to Kim Kardashian's decision to use a surrogate to carry two of her children. We each will also likely have a different take on every topic. I, for instance, could see using Trump's latest tweets to talk about what qualify as presidential statements and evidence of wrongdoing while the Kardashian piece could be a way to delve further into the topic of infertility and discuss my own struggles with IVF.

At our morning meeting we're usually joined by three top executives, the head writer, and several other producers. Each of the hosts are also joined by our personal research producers.

We're each handed a packet containing lists of the topics each of us has selected and maybe a few others, particularly if news has broken overnight. The ones that are circled are those the hosts found most intriguing. Then we just hash it out for about half an hour, deciding what we think are the best "talkers." One of the producers will usually ask what one of us generally thinks about an item, and then will turn to the group to see if anyone has an opposing opinion, because we don't want to talk about something everyone agrees on. Where's the fun or intellectual stimulation in that?

Let me pause to emphasize this: We don't rehearse or tell one another our talking points. That would be a bit of a betrayal of our audience, and we want our banter to be organic, to be spontaneous, to be authentic.

About ninety minutes before showtime, we are told which stories have made our hot topics list. The number of topics usually depends on how many guests we have that day. For instance, when Joe Biden came on after declaring his decision to run for the 2020 presidential nomination, that show was all about him. The interesting thing is that our most highly rated episodes happen to be the ones solely focused on hot topics. But *The View* has become an obligatory stop for anyone running for office, as well as the biggest figures in entertainment. We can't exactly turn them down.

We also know in advance if we're going to feature an author, and I always make sure I read the book. In fact, some publicists will actually check to make sure that I will be on set the day their author appears because they know I will have read the work and ask informed questions. To me, it's just the proper thing to do. I can't tell you how many times I've watched an interview where it's clear the interviewer hasn't read the book, or seen the play, or watched the movie. Besides being disrespectful to the artist, I also think it does a disservice to the viewer. I've got to admit that my commitment is another holdover from my school days. I'm forever the diligent student who has to do her homework. So many nights of the week, I'm at the theater, heading to the cinema, or curled up on the couch after my family has gone to bed, flipping through the final pages of a manuscript.

Whatever the subject, I approach it as if I'm preparing a legal case, which I think adds value to the show. And I also believe it forces many of our guests to make sure they're well prepared, though there are some who think they can outtalk or outsmart pretty much everyone. I will admit it, I enjoy proving them wrong.

* * *

Newt Gingrich was the leader of the so-called Republican revolution, wresting Republican control of the House of Representatives during Bill Clinton's presidency. He was hailed by many as a political genius, or at least a great tactician, until he overplayed his hand and was ousted from his role as Speaker by the same group of conservatives he had led. In the years that followed, he remained a political fixture who ran for president, appeared regularly on daily talk shows, and advised Donald Trump.

He was set to appear on *The View* a few days after Trump tripled down on his comments about there being "fine people on both sides" during the infamous tiki torch march by white supremacists in Charlottesville.

I have an amazing producer named Kristen. She doesn't have a legal background, but after three years with me, she should consider a legal career if she ever wants to take a break from television. She has an intuitive understanding of the law and is an incredible researcher who can distill information with the best of them. After our morning meetings, I can bring up any topic, say, "Kristen, you have a card on that right?" and she will. She knows how my mind works and anticipates what I'm interested in before I even ask. Some hosts rely on their producers to actually give them their points of rebuttal, but I am usually firm in my perspective. I just like to have the evidence to buttress my arguments.

Newt had been a guest at least three times since I'd been a host. And I suspected that he might try to bring to the show a canard that I'd noticed was circulating on the right, acting as though Donald Trump hadn't said that there were good people on both sides that night in Virginia when he did. I think that kind of disinformation is very dangerous and for someone like Newt to use his intellect to try to convince millions of people that they didn't hear what they heard, or see what they saw, is practically a crime. Intelligence is a gift. Education is a gift. And facts should be respected.

Trump had spoken about Charlottesville on several different occasions. The first time he said there were problems on both sides. Then a few days later, he called out the KKK and neo-Nazis, but then the next day he backtracked, saying there were fine people on both sides. Finally, right before Newt's appearance Trump praised Robert E. Lee, the slave-owning Confederate general, saying that he was great and that people at that racist demonstration were there in fact to defend his statute.

The right was trying to push this falsehood that Trump's original statement about "fine people" referenced folks who were just trying to preserve a Confederate monument, which in and of itself is worthy of discussion. But that wasn't true.

I turned to Kristen. "Please get me the transcript of each of Trump's statements. Then give me cards with the statements and dates of each."

Before the segment began, my executive producer gave me a warning. "Be very careful with Newt," he said. "He's very smart."

No longer the neophyte who didn't know what to do when her teleprompter went blank, the commentator who felt she had to mimic someone else, or the woman hoping to nab a permanent co-hosting spot, I turned to him. "He's not smarter than I am," I said. Then I headed to the set.

Sure enough Newt immediately started neutralizing Trump's statements about Charlottesville, Virginia. He said smugly, "There is this myth on the left and it's not true. Trump said clearly that he was opposed to white supremacists. He was opposed to Klansmen. He was opposed to Nazis. He says it clearly." Joy and I challenged him. "It's not that clear."

"I have what he said right here," as I pulled out my blue card and I quoted Trump's statements in chronological order. Including his now infamous "very fine people on both sides" line.

Suddenly Newt, so confident, so articulate, so impressed with his own brilliance, began to stammer. I tried to put him out of his misery. "The suggestion that he did not say that is intellectually dishonest," I said to him.

"I'm not being intellectually dishonest. I'm disagreeing with your interpretation," he said. You know your opponent's lost when he's reaching into the Kellyanne Conway grab bag of "alternative facts."

After the show, Joy sidled up beside me. "I knew you had a blue card somewhere."

I got a lot of positive feedback on social media as well. "Thank you, Auntie Sunny," someone said. "Someone brought the receipts today," exclaimed another. Yes, that was always my goal.

That's the power of this platform. That's the power of having a seat here. That's the power of being in the room.

* * *

Another time, we interviewed the rapper Meek Mill. As he discussed his years-long battle to absolve himself of a crime he says he never committed, we flashed a photo—his mug shot—showing his bruised face and one of his eyes almost swollen shut.

"How'd your eye get like that?" I asked.

He told me that I was the first person who had ever asked him what had happened to his face. That breaks my heart.

Someone without my knowledge of the justice system's inequities, someone who had not grown up seeing the trauma of poverty and police brutality, someone who hadn't been a prosecutor who tossed similar pictures back at the arresting officers, might not have bothered to ask such a question. But I couldn't sit there in front of Meek Mill, in front of our audience, and ignore evidence of apparent abuse that was right there in front of me.

Meek Mill was just one of many celebrities to come on *The View*. I'm not one to be star-struck and neither are my kids, who honestly could not care less about where I work. But the night I mentioned that Michael B. Jordan was going to be on the following day, Gabriel's ears perked up.

Jordan starred in *Creed* and *Creed II*, sequels to the *Rocky* franchise, but he's perhaps most famous for his role as the antihero Killmonger in the global phenomenon *Black Panther*. Gabriel, who's a theater fan, was not just an admirer of Jordan's acting, but was familiar with his advocacy and the fact both he and Jordan loved anime. He asked if he could leave school early to come by the studio.

Michael was a joy, and he even gave Gabriel his contact information. It was a treat that Gabriel finally, kind of appreciated what I did each day.

Another great guy was the singer Pitbull. After he came on our show, I took him aside to thank him for his work in Puerto Rico. Though he hadn't sought publicity for it, I knew that he'd sent his private plane to the island to evacuate cancer patients. I told him that I also did some work there that I didn't talk about publicly, and he told me if I ever needed him, to call him any time. When I scroll through my contacts now, I see that I've got Pitbull in my phone, which amuses me.

The great actress Glenn Close also stood out among the crowd of stars who sat at the table. She was just so normal and lovely. It's always nice when you meet people like Close, Jordan, and Pitbull, who manage to stay humble because that's certainly not always the case.

* * *

I'll always remember something Barbara Walters said.

"When people come on this show," she said, "they are a guest in your house, and you must treat them that way." Those words were so unexpected, but they made so much sense. You can disagree with the folks sitting beside you, jousting back and forth, holding their feet to the fire, but you should do so respectfully. Omarosa was a great example of someone who caused me to have to dig deep to remember my manners.

The villain on Donald Trump's first season of his reality show *The Apprentice*, Omarosa had gone on to build a career as a reality show contestant and for a while as a Trump apologist before being fired by his administration and writing a book.

When she came on to talk about her bestseller, *Unhinged*, she was incredibly rude. She came prepared for a fight. She did not speak to any of us backstage and was terribly nasty to Joy, implying that she needed to find happiness in her life. I really wanted to go low, but I didn't. After the segment, viewers tweeted up a storm, saying that I must have channeled Michelle Obama's admonition that "When they go low, we go high," which she said during the Democratic National Convention in 2016, in the midst of the ugly campaign pitting Trump against Hillary Clinton.

But I wasn't thinking of Michelle Obama. I was really channeling Barbara Walters. Omarosa was a guest in my home, and just as you sometimes have to deal gently with that obnoxious friend your husband invited over for dinner who drank too much, I had to show grace toward her. But I was glad to show her the door.

I also had some tense moments with Chris Christie, the former governor of New Jersey who later became an ABC contributor. When he and I both appeared one day on *Good Morning America*, he was very dismissive of one of my points. "Save that for *The View*," he said, cutting me off.

Christie's bad behavior was obvious enough that a network executive called me and asked me if I was all right and whether I wanted him to apologize. But I said no. I'd been around long enough to know that while I could have kept the flap going, in the end, it might hurt me more than Christie, making it appear I was too delicate to handle a bully like him.

Not long after, Christie appeared on *The View*, to promote his book *Let Me Finish*. Because I'd read it like I read the work of every author who appeared, I remarked on a passage when Christie's own sixteen-year-old son asked him if he had been behind the Bridgegate scandal that had led to a colossal traffic jam on the George Washington Bridge and the conviction of two of Christie's staffers for a petty act of political retribution. His son said to him "I've gotta ask you something. I've gotta look you in the eye and ask you, did you do it."

As a mother myself, I want my children to be proud of me. To know that I am ethical and moral. So I asked Chris, "How do you reconcile that within yourself?" I don't understand how Chris could stump for a man who made his own children question whether or not he committed a crime.

Christie was clearly annoyed. He blustered on about "politics being politics."

I didn't let him off the hook, and Christie didn't like it one bit.

He'd asked me to save it for *The View*, and I did.

Still, I'd kept it classy, something producers and many viewers have applauded me for.

What sometimes is the most unsettling is having a guest on the show, the interview is over, and they attempt to score points on social media, in my opinion because they simply can't think quickly enough on their feet or because they are too cowardly to do it on air. Likely a combination of both. That is certainly what happened when Donald Trump Jr. came on *The View* to promote his book. He came on with his girlfriend, Kimberly Guilfoyle, who I knew from my time at Fox News, although the talented lawyer I knew I no longer recognized. I asked him what I thought to be an appropriate question on air which was why he felt it appropriate to out the name of the whistleblower who alerted Congress to President Trump's abuse of power in withholding military aid from Ukraine if they did not announce an investigation of Joe Biden and his son Hunter Biden, as well as a debunked Russian allegation of Ukrainian interference in the 2016 US election. As he tried to defend himself, I explained that his actions were illegal. Kimberly, a former prosecutor, didn't disagree with my legal assessment. However, after the show, Trump Jr. cowardly tried to repeatedly attack my legal credentials on Twitter. His followers followed suit. They were the actions of a coward and also sadly, the actions of a man looking for the approval of his father. Approval he will never get. Honestly, when I looked into his eyes when he left our show, as he shook my hand and thanked me for hosting him, I saw a scared little boy. I felt pity. It was the highest rated show of the season at the time.

"Your tone was perfect," the crew will often say after we've finished taping.

"You shut him down with class," a viewer might send to me in a tweet. "That's what I love about you."

I try to behave with decorum not only because of what Barbara said but because realistically I recognize that if I get too assertive, I will instantly be branded as an angry black woman. I'll be deemed out of control, even if I'm not presenting my point any more forcefully than my co-hosts. I realize that the behavior exhibited by some of my co-hosts would find me on the unemployment line. Some of our astute viewers of color often tweet me and say that they understand why at times my speech is so clipped, slow and thoughtful. I'm a quick thinker. A fast talker. When I speak slowly, I am self-editing.

I know that I am representing an entire community and that I owe it to them to present myself, and them, in the finest way possible. I also know that if I enable others to bait me and then easily dismiss me, then I cannot be heard.

But I won't deny that all that arguing, while watching my tone and picking my battles, can be immensely stressful.

I was diagnosed with stress-related diverticulitis, an inflammation of the intestines that caused me to lose twenty pounds. I know I suffer from it because I'm a worrier by nature. I fret about my children being safe. I worry about my son having an encounter with the police. About my daughter being touched inappropriately or treated unfairly. I am anxious, given my background, about my family's financial stability suddenly withering away. And a lot of my stress flows from the show, trying to stay calm when I'm really angry, trying to be respectful when I want to tell off the rude person I'm dealing with. I have to keep it together because I know what the reaction will be if I don't.

To deal with those tensions, I turn inward. I go quiet when the chaos starts to erupt around me. At the end of the day, it's probably not healthy for me. But it's the way I've always coped. I've never been someone who sought out therapy, but I lean on my faith. My hens, clucking in my backyard soothe me; my family grounds me.

I can't say that I'm always able to shake off everything that disturbed me during the day. But at this point I'm a professional. And I'm grown enough to know you can fault people for a lack of enlightenment, for a lack of evolution, but not for their upbringing, which may have narrowed their views or embittered their feelings. I just think about how I can make my point better next time, how I can penetrate the wall someone has up when I encounter them tomorrow.

I know that for all the fans I have, cheering me on, every one of us at the table also has enemies. I get an earful from them too. But that doesn't shut me up. Every day my goal is to make an impression, to utter words that make someone watching say, "Thank God. That's just what I was thinking." Or maybe I can make someone think, "I never thought of it that way. But now, I see it."

With five hosts, you're jockeying for the space to make such breakthroughs. But if I can leave a single nugget, touch the heart of a single person, I feel I'm doing my job.

That's the thing. I think people used to dismiss *The View* as a place where five catty women fight with one another. But we are smart, we are passionate, and we discuss real issues. If you step to us, you'd better be ready to answer tough questions. In 2019, long before the first primaries and caucuses, I asked every single presidential candidate who came on our show what their opinion was about reparations for the descendants of those who'd been enslaved. The *New York Times* and the *Washington Post* were soon reporting that a question about reparations would have to be part of any debate, and it was. I know I was the first person to consistently raise that issue on a national show.

On *The View*, we tackle issues that make one another uncomfortable, like issues of privilege. When we discussed the college admissions scandal, in which wealthy parents paid for others to take exams and falsify credentials to get their kids into top schools, I was mortified and said so. "How mediocre do you have to be," I asked, "to game a system that is already disproportionately in your favor?"

Meghan and Abby, who come from very wealthy families, didn't appreciate my perspective. My impression was that they somehow felt that I may have been attacking them. Honestly, I wasn't. We had a tense exchange.

But the beauty of *The View* is that we always come back together to debate another day. The director and actor Tyler Perry actually remarked on our camaraderie after he appeared on our show. He sent each of us a lovely bouquet of flowers with a note applauding the way we had rallied around Meghan after the passing of her father, the late senator John McCain.

And in that I truly think *The View* can be a model for the nation. We tackle the most difficult topics, the ones they say you shouldn't discuss during dinner, at our table every single day. We talk about abortion, about religion, about guns, and we do it in front of more than three million more whose faces we cannot see but who definitely make their feelings known.

I realize conversations about race and class, injustice and inequality, are not easy to have, especially in the era of Donald Trump, when cruelty has become commonplace, xenophobia has crawled out from the crevices of our history, and many Americans are numbly walking on eggshells, unsure of what will come next and disbelieving that we are here, in such a moment, at all.

But I also know that no matter what policies are enacted at the Mexican border or what bans are upheld by the Supreme Court, our nation is hurtling toward a time when the majority of Americans will be people of color. We cannot afford not to talk, we cannot afford not to listen, we cannot afford to not seek common ground. Our lives, and the life of our nation, depend upon it.

I recognize that allies come from all backgrounds. I understand that sometimes people behave insensitively because they do not know any better. And I realize that a person's actions, however offensive, are not always rooted in bigotry. Just as I don't want to be labeled or written off, I know that it's wrong and unfair to jump to conclusions about the motives and mind-sets of others. When I confront a questionable situation, I conduct my own mental analysis, honed by a lifetime of interviewing witnesses, forging connections with strangers, and straddling diverse worlds. I look at the evidence to determine whether what I'm experiencing appears to come from a place of animus, ignorance, or both. And I address it accordingly.

And then there are those amazing moments when you engage with someone, a relative, a beloved friend, even a stranger, and you are able to make them look at the world in a slightly different way. Or they recognize the value of someone they had otherwise dismissed, seeing their value and promise.

Ours is a country that for centuries has been run by a minority—a cluster of privileged white men. They have hoarded their power, and some have warped the definition of what it is to be an American, making that coveted distinction one that is defined on their terms. But I believe that my family, indeed all our families, is what the American dream is truly about, a crisscrossing of economic, ethnic, and racial lines.

Even if we have to reach back to a loved one from a distant generation, at some point, each of us has been the "other." That might be easy to forget if you are not automatically viewed with suspicion when walking around a department store, or easy to dismiss if you have never had to have "the talk" with your children so that they can survive an encounter with the police, but it is the truth.

By speaking and truly listening to one another, we can begin to recognize our common humanity and tend to the diversity that has been our country's greatest strength. If we can appreciate one another fully, we can tap into all the varied talents that can make this country what it has never been but has the potential to one day be.

It's a little like what we try to do on *The View* each day, voicing our different and disparate opinions while also listening to one another. We are not always successful. Sometimes we may feel so strongly about an issue that we don't want to hear what someone who disagrees has to say. Sometimes it gets heated, or we hit an impasse and have to agree to walk away.

I don't always feel like calmly explaining my point of view because I sometimes feel that it should be so obvious what is right and what is wrong. But more often than not, even if my co-hosts and I don't find common ground, we are able to leave the set in peace. And all of us live to debate, to disagree, and to possibly reach a satisfying compromise another day.

# ACKNOWLEDGMENTS

I dedicate this book to my parents. I am because you sacrificed so much.

To my husband, Manny, thank you for being my partner in life, for holding my hand, always, especially when it's hard, and telling me that the kids would not only come, but that they would be alright.

Gabriel and Paloma, you are my blessings.

To my family and friends, thank you for letting me share our stories—warts and all.

To my teachers and mentors, thank you for the guidance.

To my literary agent Ryan Fisher-Harbage, thank you for always believing that I had a book in me. And that my story mattered. Had you not called, this would not have happened.

Charisse Jones, thank you for making my words make sense. I'm humbled by your artistry.

Thank you to my editor Hilary Swanson, and the incredible team at HarperOne. You believed that perhaps my story could lift someone up. You pushed me to write about the difficult times. And for that I am so grateful.

And to those who don't believe that their dreams are possible . . . they are.